D0045724

The Principles of Experimental Research

The Principles of Experimental Research

K. Srinagesh, Ph.D.

University of Massachusetts, Dartmouth

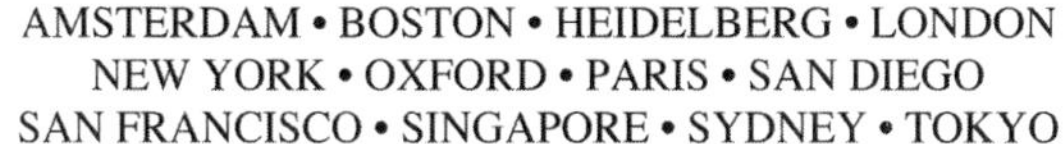

Butterworth-Heinemann is an imprint of Elsevier

Butterworth–Heinemann is an imprint of Elsevier
30 Corporate Drive, Suite 400, Burlington, MA 01803, USA
Linacre House, Jordan Hill, Oxford OX2 8DP, UK

Copyright © 2006, Elsevier Inc. All rights reserved.

No part of this publication may be reproduced, stored in a retrieval system, or transmitted in any form or by any means, electronic, mechanical, photocopying, recording, or otherwise, without the prior written permission of the publisher.

Permissions may be sought directly from Elsevier's Science & Technology Rights Department in Oxford, UK: phone: (+44) 1865 843830, fax: (+44) 1865 853333, E-mail: permissions@elsevier.com. You may also complete your request on-line via the Elsevier homepage (http://elsevier.com), by selecting "Support & Contact" then "Copyright and Permission" and then "Obtaining Permissions."

Recognizing the importance of preserving what has been written, Elsevier prints its books on acid-free paper whenever possible.

Library of Congress Cataloging-in-Publication Data
Application submitted

British Library Cataloguing-in-Publication Data
A catalogue record for this book is available from the British Library.

ISBN 13: 978-0-7506-7926-8
ISBN 10: 0-7506-7926-3

For information on all Butterworth–Heinemann publications
visit our Web site at www.books.elsevier.com

Printed in the United States of America
05 06 07 08 09 10 10 9 8 7 6 5 4 3 2 1

Dedicated
with respect and gratitude to

Ningamma and Kamataiah

of Mandya

APR 1 1 2006

Contents

Preface

I have had the opportunity of participating in many, and watching even more, research projects, ranging from industrial Research and Development (R&D) to Ph.D. dissertations. Most of these, when experimental, are still one-factor-at-a-time experiments. In spite of the fact that such luminaries as Fisher, Taguchi, and Deming have demonstrated the irrelevance of one-factor-at-a-time experiments, it seems as though these people never existed. In undergraduate education, almost as a rule, Probability and Statistics is not taught, with a conspicuous exception: in psychology. Even where "advanced mathematics" and "applied mathematics" are considered essential, for example in engineering, Probability and Statistics is ignored. This is a pitiable lack of proportion.

It is fairly common that college undergraduates have to do a capstone project in their senior year. Many of their projects involve experimental work. That is where the fixing has to be done. The students should be made aware that one-factor-at-a-time experiments are most often unjustified. To do multi-factor experiments, the students should have been taught Probability and Statistics before the senior year. Equally important for an experimenter—no matter at what level—is to be familiar with the basics of logic, for experimentation is logic in action. To realize that this is so, the experimenter should be exposed to the basics of what is known as the "Philosophy of Science."

Thus, this book is a combination of Philosophy of Science, Logic, Probability and Statistics, and Designing and Analysis of Experiments—all these aspects blended at the level suitable for upper level undergraduates or first year graduate students. It is my experience, and I submit that no undergraduate education in science—either theoretical or applied—should be considered complete without a full-fledged course on topics like those presented in this book. Even more so, it is true for students in Masters and higher level studies. And, industrial R&D researchers (I have been one for six years) who have not had a course like this before suffer from an insidious professional deficiency.

This book is not for professionals or advanced students in statistics, logic, and philosophy of science. It is for students of science, theoretical and applied. Within their domain of interest, there is need to gain a working familiarity with the above-mentioned disciplines, but there is a regrettable lack of time to study any of these in depth. Books are available in plenty: books written by statisticians for other statisticians, by logicians for other logicians, by philosophers of science for other philosophers. But such books will not serve the need of science students in colleges, or the beginners in experimental research, and hence this book. Undergraduate instructors using this book have ample scope to emphasize the topics of interest to suit the level and need of their students. More mature and senior readers should be able to read the book on their own. Selective reading from books listed in chapter-end bibliographies help them in further progress.

During the long growth period of this book, I have had the good fortune of getting generous help from many persons. My friend, Dr. Sherif El Wakil, stood by me with advice at all stages. My colleague, Prof. Noreen Cleffi read every one of these chapters, and made many corrections. Mr. Manny Pereira, Ms. Francine Gilbert and Mr. Rajendra Desai helped me in preparing the manuscript. Ms. Paige Gibbs searched several library resources for me. Mr. Joel Stein of Elsevier Science advised me in formatting the book. And, without the help of my wife, Prof. Sampurna Srinagesh, I would be still reading through mysterious words, and still searching for missed references, missed manuscripts, and missed books. I am grateful to all these people. I would be much obliged to those who find this book useful, to tell me so. Just as much, I would be obliged to those who would indicate the defects and deficiencies in this work.

1

Experimental Research in Science: Its Name and Nature

Does it contain any abstract reasoning concerning quantity or number? No. Does it contain any experimental reasoning concerning matter of fact and existence? No. Commit it to the flames, for it can contain nothing but sophistry and illusion.

—*David Hume*

Experimental research is the theme of this book. This introductory chapter attempts to delineate the scope of the book. The research referred to here is limited to that in science and its byproduct, technology. This demands a closer look at the word "science" itself, followed by the need to define the activity broadly accepted as scientific research. The so-called theoretical research in science(s) is out of our bounds. We also need to steer past quite a few activities colloquially referred to as "researching." Our concern is limited only to experimental research in science. While noting that those who do and enjoy research for its own sake are exceptions, this chapter points out some characteristic features in the working lives of career researchers, who are the rule.

1.1 Defining Science

Attempts to define the word "science" offer many variations, none of which may be complete or fully satisfactory. The word "science," a derivative from the Latin *scientia*, simply means knowledge. We know that any person with great knowledge is not necessarily a scientist, as we currently use the word. On the other hand, the word "philosophy," a derivative from Greek, means love of wisdom. When combined with the word "nature" to form "Natural Philosophy," the phrase seems to refer to the knowledge of nature; it is more specific. Until fairly recently, science was, indeed, referred to as natural philosophy. The full title of Isaac Newton's monumental work is *Mathematical Principles of Natural Philosophy.*

For reasons not easy to trace, the name natural philosophy was dropped for the preferred name, science. The intellectual distance between science and philosophy, for a time, increased, until about the early part of the twentieth century, when some of the best known scientists started philosophizing on such concepts as space, time, matter, and energy. Philosophers, in turn, found a new topic ripe with possibilities: the philosophy of science.

Returning to the phrase "natural philosophy," the word "natural" simply signifies nature. Thus, science may be understood to indicate curiosity about or knowledge, even love, of nature. If science is the study and knowledge of nature, we mean nature minus man. Man and nature are thus placed as dipoles, with man at one polarity taking a position from which he can study nature for play, curiosity, admiration, or even exploitation and gain. Nature, on the other hand, "just lies there," like an animal caged in a zoo, or worse, like a cadaver for a student's study by dissection.

As if to protest such harsh statements, in nature we have not just the inanimate part, but the animate part as well, and the above statement may be justified only for the inanimate part. The study of the animate part is broadly covered under biology with various specialties. The medical sciences, as a group, are a good example of where the polarity of man and nature gets blurred, since man himself—with life—is the subject of study, combining other sciences, such as physics, chemistry, and biology. What about technology? Much of the admiration accorded to science is derived from its accomplishments through its derivative, technology. Like a full-grown son beside his aging father, technology stands tall and broad, dependent yet defiant. With this attempt to define science broadly, we may briefly look at some definitions available:

- "Comprehension or understanding of the truths or facts of any subject" (Webster's Dictionary).

- "The progressive improvement of man's understanding of Nature" (Encyclopedia Britannica).
- "[T]he study of those judgments concerning which universal agreement can be obtained" (Norman Campbell, *What Is Science?* (New York, NY, Dover, 1953).
- "[E]ssentially a purposive continuation of . . . what I have called common knowledge, but carried out in a more systematic manner, and to a much higher degree of exactitude and refinement" (E. W. Hobson, *The Domain of Natural Science* (New York, NY, Dover, 1968).

These being only a few of the many definitions offered for science, we venture to add one more: Science is the activity directed toward a systematic search for, or confirmation of, the relations between events of nature.

1.2 Science: Play or Profession?

Apart from the definition(s), it is to be understood that science, per se, is not the theme of this book. There are many books and articles devoted to the definition, delineation, and explanation of what science is, to discussing what its aims and activities should be, and even to philosophizing about what its limitations are. The relevance, however, of science to this book is that science, unlike subjects known as "liberal arts," is very much associated with experimental research.

In terms of experiments, scientific research may be broadly classified into two categories with a slight overlap: (1) theoretical research and (2) experimental research. Some of the greatest works in physics, for example, quantum mechanics, are the outcome of theoretical research. The work of great theoretical scientists not only solve many scientific puzzles but also create new visions through which hitherto unknown relations between events can be predicted, leading to new experiments.

Several raw materials, heaped together or scattered here and there, do not make a house. It is the work of a builder to create a house out of these raw materials. In terms of science, theories are like finished houses, and experimental findings leading to some generalizations are more like house-building materials. Whereas the works of Michael Faraday are experimental, those of James Maxwell are theoretical. Some of the greatest figures in physics have been theoretical scientists: Ludwig Boltzman, Neils Bohr, Werner Heisenberg. Scientist of that caliber, who also happen to be great experimental researchers, are rather few: Isaac Newton, Errico Fermi, Henry and Lawrence Bragg. But a large number of researchers, not only in physics but in other areas of science as well, are experimental researchers.

This is not to belittle the value of experiments. In fact, no theory is valid until it passes one or more crucial tests of experiments. In engineering and technology also, some works lay claim to being theoretical; however, considering their limited domains, albeit extensive applications, they are more in the nature of generalizations based on empirical data.

As a side issue, is the work of Charles Darwin theoretical or experimental? It is true, Darwin spent time "in the field" collecting a lot of new "data" on his well-known expedition to the Gálapagos Islands. But his work, embodied in writing, consisted of fitting together the many pieces of the puzzle to form a finished picture, which came to be known as the Theory of Evolution. Experimental? Maybe. Theoretical? Yes.

This far we have looked at science as an activity. Looking at the actors in this drama known as science is even more interesting. Leonardo da Vinci is acclaimed as a great scientist; yet, his fame rests on many unfinished works. His dependence on experiment and observation rather than preconceived ideas marks him as the precursor of the experimental researcher. Curiosity led him into many fields of inquiry: mechanics, anatomy, optics, astronomy. Being primarily an artist, possibly he did not depend for his living on scientific activities.

Galileo Galilei's interest in mechanics and astronomy, Johannes Kepler's in planets, Gregor Mendel's in mutation of plant seeds, Ivan Pavlov's in conditioning dogs: all have something in common, namely curiosity to observe the way these segments of nature operate. These men, quite likely, did not earn their livings by means of their scientific interests either. It is in this sense that Ervin Schrodinger, in his book *Science, Theory and Man* (1957), equates science with the arts, dance, play—even card games, board games, dominos, and riddles—asserting that these activities are the result of surplus energy, in the same way that a dog in play is eager to catch the ball thrown by his master. "Play, art and science are the spheres of human activity where action and aim are not as a rule determined by the aims imposed by the necessities of life."

The activity of science has changed considerably since the times of Pavlov or Mendel, even since the times of Schrodinger. He writes, "What is operating here is a surplus force remaining at our disposal beyond the bare struggle for existence; art and science are thus luxuries like sport and play, a view more acceptable to the beliefs of former centuries than to the present age." In the present age, the activity of science is no more a luxury; it has become a need, though more collectively than individually. An individual, even in a scientifically advanced country, may not be cognizant of the results of science in his or her daily life; nonetheless, his way of life, even relative to bare necessities, is vastly different from that of humans even 200 years ago. The difference, more than anything else, is attributable to the fruits borne by science. The percentage of people now involved in activities that can be considered scientific is very large compared to that of 200 years ago. Further, science, which was more or less the activity of isolated, private individuals, is now more the activity of an "organization man." An individual privately working out a theory or conducting an experiment or inventing a device is a rare exception. Thomas Edison is said to have taken many patents before establishing the General Electric lab. But since his time, about a century ago, the so-called scientist now belongs to an organization. In this way, science is neither a luxury nor an activity of surplus energy. It is a full-time job, a professional career for many persons; it is no longer play.

1.3 Science and Research

The word "research," like many others, has acquired a wide currency over the past several decades. Relatively few people conducted research in the first half of the twentieth century. Those who were known to be doing research were looked upon as belonging to an elite class, often shrouded in mystery, not unlike the FBI agents portrayed in popular movies. In the imagination of the common person, researchers belonged to a secret society, the initiation into and the ideal of which were guarded secrets. Not any more. Elementary school children now ask their parents for rides to libraries or museums because they have to "do some research" on a project given by their teacher. If we need a book from the library and that particular book is not found in the list or on the rack, the librarian says, on your request for help, that she or he "needs to do some research." Finding a particular poem needs research; so does finding a music album, or a particular brand, style, or size of shoes.

With the expanding influence of consumer goods on the lives of common people, market research has acquired great power. Firstly, the products that members of society consume, be they houses, cars, items of clothing, or sunglasses, are the outcomes of research. Secondly, the various subtle and sophisticated processes of persuasion—the brand names by which a product is called, the faces that flash, the music that plays during a commercial, the pictures of heroes, stars, or muscle men on packaging—are all subject to market research.

The service industry is just as much shaped and controlled by research. The kind of plays, movies, or TV shows that are likely to become popular (hence, profitable) are not guessed and gambled on any more. Entrepreneurs intending to start a new product a few decades ago needed to do the familiar "market survey." Now, they go to specialty companies that "do research" to find the answer to fit the client's requirement. Lawyers searching for precedents, doctors looking for case histories, accountants looking for loopholes to minimize taxes: all engage in matters of research. Though one may question the accuracy of the word "research" in these contexts, the word, of course, is free for all. But we want to point out that research, as discussed in this book, is meant to be scientific research in its broadest sense. Accordingly, ornithologists observing the nesting habits of the peregrine falcon, pharmacologists trying to reduce the side effects of a drug, zoologists planning to clone a camel: all these, besides hundreds of other activities, may be considered academic, technical, or scientific research.

Further effort to delineate the scope of this book calls for a bifurcation, somewhat overlapping, between research activities that are scientific in nature and those that are not; this book deals only with the former, meaning, research in those areas that are conventionally considered "science." Thus, although several thousand Ph.D. dissertations being written throughout the world in the areas of philosophy, political science, literature, and so forth, are research efforts, these, for our purpose, offer marginal interest. And even within science, this book deals only with those research works that are experimental in nature. This distinction requires that we clarify the phrase "experimental research" even further.

1.4 Varieties of Experimental Research

To experiment is to try, to look for, to confirm. A familiar example may make the meaning clear. After getting ready in the morning to go to the office, I look for the keys. I can't find them. I am mildly agitated. Thoughts of not getting to the office, of not getting the notes I wrote yesterday, of going late or empty-handed to class, and so forth, run quickly through my mind. But I know the first thing I need to do is look for my keys. I run up again to my bedroom and look in the same chest of drawers where I keep them every night. Not there. I look in the next drawer. Not there either. Then, I recollect that I was watching TV last night. I go into the TV room and look on the TV table and places near about. Not there. Then, I think of the clothes I wore yesterday. I go to the closet, recollect what I wore, and search the pockets. Not there. Did I, by any chance, leave them in the car when I got out of it yesterday? I look through the locked door of the car. Not there. Then, I remember that I needed the keys to lock the door; so, I decide, because the doors are locked now, there is no chance that the keys are inside the car. I go back into the house and ask others in the family if any of them have seen my keys. They start, half-heartedly, looking for them. In the meanwhile, I go up to the bedroom, rehearse in my mind—even partly, physically and orally—how I came into the house yesterday, where and how I keep my briefcase, where and how I removed my jacket and shoes, how I was called to eat dinner soon after, how I did not fully change my clothes before going to the table, and so forth. I go through the places involved, look at all surfaces on which I could have placed the keys, step by step in sequence. I even open the refrigerator and look inside—just in case. My frustration, of course, is mounting steadily.

At this stage, the reader may object that this is all searching; where is the experiment? Indeed, it is searching, but searching with a difference. The difference is that the search is not haphazard; it is not arbitrary. Frantic though it appears, it is systematic; it is organized. It was done with one question in mind all the time: Can my keys be here? The search is not extended everywhere: to the bathrooms, the attic, the basement, the backyard, or even to other bedrooms. It is done only in places where the keys are likely to be. The circumstances that could cause them to be in selected places are thought about. To use a more precise language, it is a search directed by one or more hypotheses.

If my boy wants to collect seashells, I will take him to the seashore, not to the zoo. If he wants to collect pinecones, I advise him to wait until late fall, then to go and look under pine trees, not under any tree in the park. Every parent, with some sense, does the same. Primitive as they seem, these are logical decisions: to search with a question in mind and to do so at a time and place and under the circumstances likely to yield a favorable answer.

I am aware that these examples are not adequate in strength to reach the textbook definition of "experiment," but these are experiments, nonetheless. Very famous scientific works, like the discovery of the planet Neptune in 1846, which was hailed as one of the greatest achievements of mathematical astronomy, notwithstanding all the work it involved in mathematics and astronomy, is eventually a search. Many theoretical calculations were obviously needed to direct the search at a particular location in the vast sky at a particular time. But is it far-fetched to call the search for the object itself an experiment?

If the above is an example of a deliberate search, backed by an expectation, there are instances in the world of science in which unexpected happenings led to discoveries of monumental scale. Wilhelm Roentgen's discovery of X-rays and Alexander Fleming's discovery of antibiotics are, to outsiders, just accidents. But such accidents can happen only to those who are prepared to perceive things that most of us do not. The culture of the mind of those scientists was such that they could decipher meaning in these "accidents"; they could expect answers to questions they were prepared to ask, even if they had not consciously asked them yet. The preparation of their minds embodies the question; the accidents, the answers. The accidental discoveries, subject to further confirmation or variation, become part of science.

The training of athletes is another example of experiment. Several timed runs of a sprinter, several shots to the hoop of the basketball player, under repetitive conditions and often with intentional variations, are answers to the combined questions, Can I do it again? Can I do it better? The housewife's cookies, the chef's old and new dishes, the old and new styles of dress items, new creations in art, music, and literature: all are, in a sense, experiments backed by hypotheses. Logic is implied, if not expressed, in each. That the word "research" ends with "search" is, quite likely, purposive.

Even the zest of the gambler—with various games, combinations, and numbers—is not devoid of logic, though often, when it comes to detail and when it is too late, he recognizes that logic worked, not for, but against him.

1.5 Conventional Researchers

The most typical and common experimental works of our times, are the works of

- Research scientists and technologists employed in a career in private or government-supported institutions and organizations, including the military
- Thousands of scientists and engineers in the making, namely registered graduate students working toward university degrees, such as an M.S. or Ph.D.
- Thousands of inventors, some affiliated with organizations and some private, working for patent rights on materials, processes, and devices

State governments in which a ministry (or department) of science and technology does not exist are exceptions. Most countries spend a considerable part of their revenue on research work. Almost all countries have military establishments, a considerable part of whose resources are directed to defense research.

The following are some common features among researchers in such a wide variety of research fields:

- Most of them work either individually or in groups, under the supervision of a guide or a director.
- They work for a short or long term or for promotion in the form of a degree or other enhancement of professional status.
- They have the intellectual support, the library, and a field of work, the laboratory.
- Last, but by no means least, they have "the problems." Their efforts are directed toward finding answers to questions posed by individuals or organizations.

This last aspect, namely, the existence of questions and the efforts toward answers, is to us the significant part of experimental research.

I have had the unpleasant experience of having registered in a prestigious institution to do research for my Ph.D. without having identified a research problem. And this, as I now understand, is not a rare incident, particularly in developing countries. That institution, at that time, was devoted to research, as well as to teaching engineering at the graduate level, but it was devoted to research only in the physical and biological sciences. Students were registered for research positions, mostly for Ph.D's, the main attraction being the availability of federal funds for scholarships and research assistantships. The institution requisitioned a yearly grant and was awarded a certain percentage of requested, based on the prestige and politics exercised by the people involved.

I cite this example to illustrate the situation of putting the "Ph.D. before the problem," the cart before the horse, in that the student, the would-be Ph.D., knows that he is working for a doctorate, but he doesn't know what his topic of investigation, his "problem," will be. In my case, the situation continued for more than a year and was likely to continue indefinitely. Luckily for me, I came up with a problem for myself after a year or so. That situation is far from ideal. I invented a problem through literature search not because the solution to that problem was a pressing need, but solely for the purpose of getting a Ph.D.

In an ideal situation, on the other hand, the "problem," at least in outline, should be available, and a certain amount of funding should be earmarked in the budget for the effort of finding the solution. That effort should be assigned to a suitable student, with compensation coming from the funds. A healthy and desirable situation exists when the problem(s), along with the funds come from those who face the problem, whether it is an agency of the local or federal government or a manufacturing (or service) company. This is what, in principle, is happening in developed countries, for example, in the United States. In developing countries, in view of the limited availability of funds, it is even more desirable that this model should prevail.

This brings us to the last and most idealistic kind of research: research done in the same spirit as playing a game, creating a piece of music, or painting a picture, not for gain of any kind, but simply out of exuberance and energy, just for enjoyment, for its own sake. Certainly such a situation is very desirable to the cause of science, as well as to the enjoyment of the people involved. There have been many such people in earlier times whose contributions to science have been significant, but they are now a vanishing species, if not endangered as well.

1.6 Bibliography

Campbell, Norman. *What Is Science?* Mineola, NY: Dover Publications, 1953.

Conant, James B. *On Understanding Science*. New York, NY: New American Library of World Literature, 1955.

Hobson, E. W. *The Domain of Natural Science.* Mineola, NY: Dover Publications, 1968.

Hutten, E. W. *The Organs of Science.* New South Wales, Australia: George Allen and Unwin, 1962.

Schrodinger, Erwin C. *Science, Theory and Man*. Mineola, NY: Dover Publications, 1957.

Part I: The Fundamentals

2

The Importance of Definitions

Language is the principal tool with which we communicate; but when words are used carelessly or mistakenly, what was intended to advance mutual understanding may in fact hinder it; our instrument becomes our burden.

—Irving M. Copy and Carl Cohen

The need for definitions in experimental research emanates from the fact that experimental researchers in a given domain of nature are spread out widely over space and time. Ideally, there would be no second-guessing among them on the meaning of a word or combination of words. We start with a famous dialog attributed to Socrates, which emphasizes the need for definitions. With that backdrop, this chapter attempts to study the structures and functions of definitions. Differing situations require different forms of definitions; some of the more significant ones are discussed. A few informal rules as to what definitions should and should not do are presented.

2.1 Toward Definition

The following is a segment from the *Dialogues of Plato*[1]:

Socrates: "What I asked you, my friend, was, What is piety? and you have not explained to me, to my satisfaction. You only tell me that what you are doing now, namely, prosecuting your father for murder, is a pious act."

After some cajoling from Socrates, Euthyphro offers this definition:

"What is pleasing to the gods is pious, and what is not pleasing to them is impious."

Socrates effectively shows Euthyphro—in fact he gets a "confession" from him—that gods may differ among themselves, may even quarrel, with some gods accepting certain acts as pious and others coming hard on the same act as impious. So bringing "gods" into the definition does not help in defining piety. The dialog continues,

Socrates: "[P]iety and impiety cannot be defined in that way; for we have seen that what is displeasing to the gods is also pleasing to them. So I will let you off in this point Euthyphro; and all the gods shall agree in thinking your father's deed wrong and in hating it, if you like. But shall we correct our definition and say that whatever *all the gods* hate is impious and whatever *they all* love is pious: while whatever some of them love and others hate, is either both or neither? Do you wish us now to define piety or impiety in this manner?"

Euthyphro: "Why not Socrates?"

Socrates: "There is no reason why I should not Euthyphro. It is for you to consider whether that definition will help you to teach me what you promised."

Euthyphro: "Well I should say that piety is what "*all the gods* love and that impiety is what *they all* hate." [emphasis mine]

The dialog, of course, goes further. There are two points worthy of note. Firstly, Socrates insists on defining a quality, piety, so that in any further discussion, he and Euthyphro shall have a common ground. And secondly, he is, in fact, defining what a definition should be. If *some* gods agree that a certain act is pious, and *some other gods* disagree, holding the opinion that the same act is impious, then the word "piety," based on that act, is not defined satisfactorily. If, instead, *all the gods* agree that a certain act is pious, then that act serves the purpose of being an example, and the word "piety" can be attached to it. So, a definition should be such that it helps avoid disputes or disagreements. In the situation quoted, it is the dispute, the disagreements among gods, that is evidenced. In our context, in place of gods for the ancient Greeks, we find other researchers, each of whom is a self-appointed judge and the harbinger of truth in his little corner of the world. And to make the situation worse, these earthly gods are usually scattered over wide stretches of space and time. Any word or combination of words, including symbols, that a researcher employs in his discourse, needs to be clear beyond the possibility of being mistaken by any other researchers in his field; that is how definitions become relevant.

Definition is the domain of logicians and philosophers. But it is necessary that the experimental researcher should have some familiarity to build on, in case he needs either to use definitions in his own work or to understand definitions in the works of others. To that extent, we will review, in brief outline, some important aspects of definitions. In addition, this chapter aims to impress upon the experimental researcher that the process of defining, when required, should not be done casually.

2.2 Defining "Definition"

How do we define "definition"? It is somewhat disappointing to learn that there is no one definition, acceptable to most, if not all, logicians. This surprise will soon disappear when we notice that several functions performed as a way of clarifying expressions (or statements), either completely with words or with combinations of words and symbols, have all been considered legitimate acts of defining. Consequently, there are many variations of the meaning of "definition."

Having accepted this situation, we will mention (rather, quote somewhat freely) a few variations, meant to give the breadth of the word "definition," though not its depth.

(A) A real definition involves two sets of expressions, each with a meaning of its own, and these meanings are equivalent if the definition is true.[2]

(B) A definition is a phrase signifying a thing's essence. (Furthermore,) A definition contains two terms as components, the *genus* and the *differentia*. A genus is what is predicated in the category of essence of a number of things exhibiting differences in kind, and the *differentia* is that part of the essence that distinguishes the species from the other species in the same *genus.*

—Attributed to Aristotle[3]

Four of the following definitions, (C) through (F), appear within a span of six pages in one book.[4] The rest are taken from other sources.

(C) A definition is a way of restricting the meaning of a term that has several meanings to prevent ambiguity and equivocation.

(D) A definition is true . . . if the defining term has the same designation as the term to be defined.

(E) Definition is a rule of substitution.

(F) "Definition" = statement that one term has the same designation as another term.

(G) A definition is the statement of the meaning of a word, etc.

(H) A definition is a brief description of a thing by its properties.

(I) A definition is an explanation of the meaning of a word or expression.

2.3 Common Terms Used in Definitions

Though the word "definition" itself defies definition, the components that constitute a definition are delimited fairly well as follows:

Term: Word or group of words, such as phrases, clauses, or one or more assertive sentences. To the extent that terms are defined, when need be, communication between or among researchers becomes precise. In the Socratic dialog cited earlier, the term to be defined was the word "piety." In dictionaries or books of synonyms, we may find several alternate meanings for a given word to be defined. Out of these alternatives, the one that best explains the term and—this is important—best fits the context may serve the purpose of being the definition of the word. There may be occasions when the term is a group of related words, a *phrase,* wherein it is not the literal meaning of individual words that is significant, but that of the bunch of words as a whole, such as "shadow boxing," "end of the tunnel"; these are often referred to as *idioms*. In such cases, we need to go to books of idioms, which, though not as common as dictionaries, are available in most well-developed languages like English. The language of science is usually sober; there is little room for idioms like those above or for other figurative words or word combinations.

The need to define a term when the term is a whole sentence is less often encountered; when it is, the definition usually takes the form of an explanation, often with illustrative example(s). It is worthy of note that in his great work, the *Principia*, Isaac Newton starts with definitions, covering the first dozen or so pages, before even mentioning the problems addressed therein. One of the twelve definitions, as an example, is quoted below:

"Definition VII: The accelerative quantity of a centripetal force is the measure of the same, proportional to the velocity which it generates in a given time.

Thus the force of the same loadstone is greater at a less distance, and less at a greater: also the force of gravity is greater in valleys, less on tops of exceeding high mountains; and yet less (as shall hereafter be shown), at greater distances from the body of the earth; but at equal distances, it is the same everywhere; because (taking away, or allowing for, the resistance of the air), it equally accelerates all falling bodies, whether heavy or light, great or small."[5]

***Definiendum and Definiens*:** These names are given, respectively, to the term to be defined and the term that does the defining. One or both of these terms may be one or more words, or even, as we have just seen Newton do, one or more sentences.

2.4 Varieties of Definitions

2.4.1 A. Direct and B. Indirect Definitions

Direct definitions are explicit in nature; hence, the *definiens* can replace the *definiendum* without any further need for elaboration or explanation. If a definition forms a part of a whole statement, and if after replacing the definiendum with the definiens, the statement can be repeated without any loss or alteration in the original meaning, it is also a case of direct definition. In contrast, *indirect definitions* are such that by replacing the definiendum with definiens, both of these being either isolated or part of a statement, the meaning of the statement remains open to further relevant questions. There are two variations within this. Firstly, when a word or a combination of words conveys meaning far beyond what a usual-length definiens can clarify, because the definiendum may have several aspects, some of which are implied and cannot be demonstrated, the definition is referred to as an *implied definition*: "religion," "democracy," and "honesty" are some examples. It is often the case that such definitions have emotional overtones. Secondly, if *y* is the descendent (the word to be defined) of *x*, then *y* may be a son, a grandson, or many more generations removed, and yet be the descendent of *x*. The definition then for "descendent," the definiendum, is uncertain and open to further question, in this case, as to how many generations removed or recurring. The definition, whichever way it is offered, needs to be qualified; this is often referred to as a *recursive definition*.

2.4.2 C. Informal and D. Formal Definitions

In most cases of human discourse, definitions are blended so nicely that we do not notice them as such. In a sense, every word of every language, either spoken or written, may be considered a definition. We live with these without needing to be conscious of their definitional nature. Most experimental scientists, most of the time, enjoy the same privilege. But occasions may arise unnoticed, though rare, when additional effort may be necessary to highlight the aspect of "definition" in their discourse. The degree of highlighting required and the amount of clarity intended, among other circumstances, decide the degree of formality that is desirable in the process of defining. Defining done with a low degree of formality is usually referred to as *informal definition*. Suppose I were writing for a tennis magazine on the advantages and disadvantages of "bubbleballing." I might write something like this: "Some players are likely to return the balls to the opponent intentionally and repeatedly, hitting the ball high above the net, making the ball drop to their opponent almost vertically near the baseline. For our purpose we may call this 'bubbleballing.' Here are some advantages of bubbleballing." I would proceed to write on, using bubbleballing instead of the rather long definiendum mentioned above. This is an example of informal definition.

Informal definitions can be stated in several different ways, using different sets of words; a few variations follow:

1. Bubbleballing is the act of returning balls, the path of which to the opponent resembles a bubble.
2. The word "bubbleballing" is applied to the way a tennis player returns the ball to his opponent with a big upswing, followed by a big downswing.
3. A player who does "bubbleballing" is understood to be returning the ball to his opponent in tennis, deliberately hit high into the sky.

So long as the sense in the expression is conveyed, some residual vagueness or ambiguity is not frowned upon. However, we require *formal definition* most often in research, and it needs to be done with a tighter grip on the words. One possible way is, "'Brainwashing' has the same designation as 'Changing the other person's opinion by subtle repetitions of slogans.'" Even more formal definitions avoid the words altogether between the definiendum (x) and the definiens (y) and connect the two with the "=" sign in the form "$x = y$"; the "=" does not have the same meaning as in mathematics. Originating from *Principia Mathematica* by Alfred Whitehead and Bertrand Russell, a formal way of defining has come to be widely accepted. It has the following form:

Beauty . = . that which is pleasing to look at. Df.

The term on the left-hand side is the definiendum and that on the right-hand side is the definiens, ending with "Df." to denote that this is a definition.

2.4.3 E. Lexical and F. Stipulated Definitions

Lexical definitions, obviously, are the meanings as listed in the dictionaries. As such, we find the current and established meaning(s) of a word. For instance, in the United States, currently, the word "suck" has acquired a meaning that has sexual connotation, unlike in other English-speaking societies or in the past. It is reasonable that in the near future, we will see this new usage reflected in American dictionaries. Also, dictionaries list more than one meaning for many words. It is then left to the individual to find the appropriate meaning by context.

Stipulated definitions assign special or restrictive meanings to words (or combinations of words) that otherwise have a colloquial usage, which is most often obvious. "Stress," for instance, is a word commonly used to connote that someone is mentally tired, but engineers take the same word, define it as "load per unit area," and assign to it a mathematical format:

$$\sigma = P \div A$$

where σ stands for stress, P for load, and A for area.

A variation of the stipulated definition will occur when a word (or a combination of words) is improvised to describe a certain thing or phenomenon within a limited and exclusive domain. The use of the word "bubbleballing" within the game of tennis is an example. In such circumstances, the definition is known as an *impromptu definition*. It is obvious that such definitions should not have currency outside the particular domain, in this case, the game of tennis.

2.4.4 G. Nominal and H. Real Definitions

A *nominal definition* is most often a common understanding of what a certain word or group of words should mean for the users. In this sense, the dictionary meanings of words in any language have this characteristic. The entire human discourse depends on words, though we seldom have occasions to notice these as definitions. In mathematics and the sciences, we depend on a large number of symbols. That "3" stands for "three" and "23" stands for "the sum of two tens and three ones" is so much a part of our routine that we do not think of them as definitions. In addition to the economy of space and time, both in writing and reading, such symbols are instrumental for the clear thinking needed for further development. What distinguishes a nominal definition is that it is neither true nor false and, hence, cannot be a proposition. No Briton can charge an American with being "untrue" if the latter writes "center" where it ought to be, according to the Briton, "centre." The same is true for symbols; for instance, the current symbol for gold (in chemistry) is "Au." If there is a move in some future time among chemists, who agree to do so, it may be changed to "Go," or any other letter or group of letters, without rendering any statement made thus far "false."

In contrast, a *real definition* can serve as a proposition, which means that it is either true or false, not by itself, but as decided by individual people. If "music" is defined as "a series of sound variations, capable of producing a pleasing sensation," then there is plenty of room to dispute whether some of modern music is music or something else, to be defined using different words as definiens.

2.4.5 J. Definitions by Denotation

Denotation is a way of further clarifying the meaning of a term by means of examples or instances, which most often follow, but may precede, the formal part of the definition. A good example is Newton's definition that we quoted earlier, wherein the passage "the force of the same loadstone is greater at less distance" is used to substantiate "the accelerative quantity of a centripetal force" that he is defining.

2.4.6 K. Ostensive Definitions

Ostensive definitions cannot be described exhaustively by words alone but can be demonstrated or pointed to easily to obtain complete satisfaction. If a painter is asked to describe (or define) yellow ochre as a color, the one way most suitable to him is to squeeze on his palette a thread of paint from his tube of yellow ochre and ask the other person to look at it.

2.4.7 L. Definitions by Genus and Difference

Attributed to Aristotle, such a method of definition depends on showing a particular entity as belonging to a set and nonetheless being different from all other elements of the set. The famous example is "Man is a rational animal"; in all respects, man is another animal, with the difference that he alone is endowed with rationality.

2.5 Need for Definitions

We opened this chapter quoting the dialog of Socrates on definition as a way of showing how definitions are created. The need for definition, according to many modern logicians, arises because of the possible effects of what are meant by the two words most often mentioned: "vagueness" and "ambiguity."

"Vagueness" itself may be understood as "the quality of not being clearly stated (or understood)," and *ambiguity* as "the quality of an expression whose meaning can be taken in two or more ways." No language is perfectly free of vagueness or ambiguity; English is no exception. An experimental scientist, in his profession, needs to express his findings, be it for the benefit or the judgment of others. His expressions, as interpreted by others, should have a one-to-one correspondence with what he meant to express. Ideally, there should be no occasion for him to say, "No, that is not what I meant"; there should be no such excuse because, as we pointed out earlier, many of his potential judges or benefactors, as the case may be, will be scattered far and wide, in both space and time. Slackness in definitions may lead researchers elsewhere, or those yet to come, into mistaken tracks of investigation.

Fortunately for the experimental researcher, most of his or her work can be expressed with close-to-ordinary words and symbols, there being less need for strict and formal definitions. But when the need arises, definitions should be considered as tools, and if the researcher is not familiar with the use of these, his or her job may become difficult, if not impossible.

The motivation for definition can also be either the need for economy of words, or contrarily, the need for elaboration or clarification as shown following:

Abbreviation:

My father's father's other son's son . = . my cousin. Df.

Elaboration:

Here is an instance of ambiguity. My cousin could be either:

my father's brother's son or daughter, or

my father's sister's son or daughter, or

my mother's brother's son or daughter, or

my mother's sister's son or daughter. In way of specifying:

My (this) cousin . = . My father's father's other son's son. Df.

2.6 What Definitions Should and Should Not Do

Even among logicians, there is no unanimity as to what ought to be called "definitions." Having said this, we mention below briefly, without the constraints of quotation, some of the desirable and some of the undesirable traits of definitions, as expressed by logicians.

A definition should

1. Make communication possible when it is impossible without it, or make communication clear when it would be fuzzy without it

2. Have two terms: (a) the term to be defined (the meaning of which, in the context, is doubtful), and (b) the term that does the defining (the meaning of which is expected to be understood)

   Example: Painter . = . one who paints pictures. Df.

3. Distinguish between things and words

   Example: In "Anger is harmful," we are talking about the thing (emotion) "anger." In contrast, in "'Anger' has five letters," we are talking about the word "anger" (not about the emotion "anger").

   The means of making this distinction is to use the quotation marks judiciously.

4. Distinguish between the noun and verb forms of some words, which can be used in both forms.

Example: "I am *writing* this passage," versus "This *writing* is done well."

5. Give the essence of that which is to be defined. The definiens must be equivalent to the definiendum—it must be applicable to everything of which the definiendum can be predicated, and applicable to nothing else.

6. Be so selected that, whether explicit or implicit, the attributes known to belong to the thing defined must be formally derivable from the definition.

A definition should not

1. Use examples as the sole means of defining, though examples may supplement a given definition. We have seen this done, as it should be, in the definition quoted from Newton's *Principia*.

2. Use description as the sole means of defining. Here again, the definition quoted from Newton's *Principia*, done as it should be, may be considered as containing a supplementary description.

3. Use exaggeration (as a form of definition)

   Example: "Definition" by Bernard Shaw:

   Teacher: *He who can, does. He who cannot, teaches.*

4. Be circular; it must not, directly or indirectly, contain the subject to be defined (some times referred to as *tautology*)

   Examples:

   a. Hinduism . = . the religion followed by the Hindus. Df.

   This is obviously and completely circular.

   b. Hinduism . = . the religious and social doctrines and rites of the Hindus. Df.

   This is from a respectable dictionary; the circularity is obvious, though not direct.

   c. Hinduism . = . the religious and social system of the Hindus, a development of ancient Brahmanism. Df.

This is from another respectable dictionary. The addition of the phrase "a development of ancient Brahmanism" is an improvement, but not in the direction of reducing the circularity.

Instead, use the form:

d. Hinduism . = . religious and social rites and doctrines that are a development of ancient Brahmanism. Df.

The circularity is completely avoided, though the new word introduced, "Brahmanism," needs to be defined, in turn.

5. Be phrased in the negative when it can be phrased in the positive

Example:

a. Night . = . part of the day wherein there is no sunlight. Df.

b. Night . = . the time from sunset to sunrise. Df.

Though (a) may be literally correct, (b) fulfills the logical requirement better.

However, there are legitimate exceptions.

Example:

c. Orphan . = . child who has no parents. Df.

This is acceptable, though it is possible to remove "no" by defining the word differently as

d. Orphan . = . one who is bereaved of parents. Df.

6. Contain obscure term(s)

This pertains to the purpose of definition, namely, to clarify, not to complicate, confuse, or lead astray. The earlier example, in which "Brahmanism" was used to define "Hinduism," is an instance.

2.7 References

1. Plato, Euthyphro, Ethics: *Selection from Classical and Contemporary Writers* Ed. Oliver A Johnson, 6th ed. (Orlando, FL: Harcourt Brace Javanovich, 1989) 16–26.

2. M. R. Cohen and E. Nagel, *An Introduction to Logic and Scientific Method* (London: Rutledge and Kegan Paul, 1961), 238–41.

3. Ibid, p235.

4. M. C. Beardsley, *Thinking Straight* (Upper Saddle River, NJ: Prentice Hall, 1951), 159–65.

5. Isaac Newton, *Sir Isaac Newton's Mathematical Principles of Natural Philosophy and His System of the World*, Motte's Translation, 1729 (Berkeley: University of California Press, 1974), 1:4.

2.8 Bibliography

Copi, Irving M., and Carl Cohen. *Introduction to Logic*. 8th ed. New York: Macmillan Publications Co., 1990.

Fujii, John N. *An Introduction to Elements of Mathematics*. John New York: Wiley and Sons, 1963.

Russell, Bartrand. *Principles of Mathematics*. 2nd ed. New York: Norton Company, 1964.

Whitehead, Alfred N., and Bertrand Russell. *Principia Mathematica* 2nd ed. Cambridge Univiserty Press, 1925.

3

Aspects of Quantification

When you can measure what you are speaking about and express it in numbers, you know something about it, but when you cannot measure it, when you cannot express it in numbers, your knowledge is of a meagre and unsatisfactory kind.

—*Lord Kelvin*

Measuring parameters, independent and dependent, forms the functional basis of most experimental research. Starting with the use of numbers, the principles of measurement are briefly analyzed. Quantities arise in the process of measuring and denoting the degree or intensity of qualities, and the measurement is made with *units*. This chapter deals with the uses of units and the need for dimensional analysis when multiple units have to be used in combination.

3.1 Quantity and Quality

One can think of several cases in which the distinctive feature that separates the scientific from the nonscientific is *quantification*, which means expressing laws or relations in terms of quantities combined with qualities rather than by qualities alone. To mention that New York city is full of criminals is a nonscientific expression in contrast to providing the data about how many criminals, decided by such and such criteria, are in the custody of the New York City Police Department. In the latter case, two items render the statement scientific:

1. The criteria needed to decide how to identify the criminals
2. The enumeration of those who fulfill such criteria

The first of these is a matter of definition (see Chapter 2); the second, namely, the enumeration involved, is a matter of quantification. Even in terms of our previous discussion relative to the function of science, we may think of the relation between those identified as criminals and the number of those who constitute such a group as a *law*. In such statements, which are quantitative and thereby can be claimed to be scientific, the relation is not between two events; it is between a certain defined quality and the number of those who have such quality. In such statements, there is need for a definition and to count those entities that fulfill that definition. There is no room left to make arbitrary, biased, and unconfirmed statements. If it is one's hypothesis to brand New York as a place full of criminals, the hypothesis needs to be confirmed by experimentation, in this case done by checking the definition of "criminal" and further by enumeration of the criminals. The process of enumeration, better known as counting, is done by means of numbers.

3.2 The Uses of Numbers

Recognition of the importance of numbers, the basis of all quantification, dates back to Pythagoras (572–500 BC). One of the first steps ever taken toward the formation of science was counting, using numbers. Otherwise, it is hardly possible to express anything with at least some degree of the precision needed for science. Whether the statement is about an extensive quality, like the "hugeness" of a building, or an intensive quality, like the "hotness" of a summer day, we need to count by numbers.

But numbers are known to serve many purposes. Firstly, numbers are used as tags, as in the case of the route of a city transport bus or house numbers on a street. Numbers may also stand for names, as in the case of patients in a hospital ward or a private in a military regiment. In both these cases, the numbers have no significant quantitative values. Route 61 may have no relation whatsoever relative to any conceivable quantity. The patient in bed number seven may very well be treated without the hospital staff knowing his name. He can be moved to another bed and given another number, which becomes his new "name." Such numbers have, besides identification, no significance in science. Then, there is the "ordinal" use of numbers, which denotes the order of a certain attribute or quantity or even a quality, but without quantitative values. For instance, if three cities in a state are numbered according to their size based on population, city number two may be smaller than city number one by a million, and city number three may be smaller than city number two by only a few hundred. Here the significant relation is (in terms of population)

City number 1 > city number 2 > city number 3

If another city's population, hitherto unknown, is newly determined, it is possible to place it in the appropriate slot relative to the three cities. If this idea is extended to a large group of items, into which hitherto unknown items, newly found, need to be added, any new item can be assigned its appropriate relative spot in the series of all items of its kind. This is an important relation often used in science, as in the case of the identification of elements in the periodic table. Another important relation made more explicit in mathematics is that

If $A > B$, and

$B > C$, then

$A > C$

This law, known as *transitivity*, is often used in scientific relations. The ordinal numbers, though related to quantities, as above, do not directly denote quantities. For instance, consider a series of twenty different solids in order of increasing density as *S1, S2, S3, . . . S20.* In this, *S18* is known to be denser than *S1, S2, S3, . . . S17* and to be less dense than *S19* and *S20.* It does not imply, however, that *S18* is two times as dense as *S9*, the "two times" coming from dividing 18 by 9.

Such denotative significance is assigned to the so-called cardinal numbers, which are meant to answer the question, How many?

3.3 An Intellectual Close-up of Counting

Counting requires that the group within which it is done be separable. If it is said that there are more people in Chicago than in New York, it is understood that the areas that are officially demarcated as Chicago and New York are known without ambiguity, relative to their various suburbs, that only people living within the respective areas are counted, and that the number of people so counted within Chicago is more than the number of people so counted in New York. An assumption made is that the entities counted—in this case, men, women, and children—are discrete. (A pregnant woman due to deliver a baby in a moment is counted as one, not two.)

The theory of numbers involved in counting, done by rote by most civilized persons, has been found worthy of analysis by some of the greatest mathematicians. Suffice it for our purpose to record the three basic rules involved:

1. If a set A of discrete entities and another set B of discrete entities are both counted against a third set C of discrete entities, then if A and B are counted against each other, they will be found to have the same number. We may formalize this relation (mathematically) as

   If $A = C$ and $B = C$, then $A = B$.

2. Starting with a single entity and adding continually to it another entity, one can build up a series (or set) of entities that will have the same number as any other collection whatsoever.

3. If A and X are two collections that have the same number, and B and Y are two other collections that have the same number, then a collection obtained by adding A to B will have the same

number as the number obtained by adding X to Y. In terms of mathematics, we may state this as

$$\text{If } A = X \text{ and } B = Y, \text{ then } A + B = X + Y$$

These apparently simple, to most of us obvious, rules characterize the cardinal use of numbers, which form the basis of counting; particularly familiar is rule 3. Let us say, for some purpose, that the total population of Chicago and New York City together needs to be measured. To do it, we do not require that all men, women, and children of both these cities be huddled together in a place where we do the counting from start to finish. We do the following:

1. The population of Chicago is found; it is noted as a number, say Nc.
2. The population of New York City is found; it is noted as a number, say Nn.
3. The population of Chicago + the population of New York City is found as $Nc + Nn$.

This very simple procedure, obvious to most of us, is an application of rule 3 of counting above. Though the other two rules are even more fundamental, this one, by virtue of being "mathematical," presents explicitly the fundamental principle of the use of cardinal numbers, namely, counting.

3.4 The Process of Measurement

After admitting that counting is the basis of measurement, we may ask the question, Can any quality whatsoever of a thing or things (external to us) be measured? If we supply the answer no to open the issue, then a more specific question is, What are the qualities that can be measured? Let us say that a man has a bunch of red roses in his hand. We can readily think of two qualities: (1) all the many subqualities like the color, smell, shape, and structure that characterize the familiar red rose, and (2) the fact that there are twelve of them, each of which is interchangeable with any other in the bunch. The "twelve" in this is as much a quality of the bunch as the first and is symbolized by the numeral "12." Now, let us say that the person holding the red roses decides to buy another bunch of red roses. When he then holds the bunches together, there occurs no change in the quality "red roses," but a change does occur in the quality "twelve." If the person now holding the red roses wants to "measure" this new quality, he counts the total number of roses. Needless to say, he may do it in several different ways.

One-by-one, arithmetically symbolized as

1 + 1 + 1 + . . . until he counts the last one

Two-by-two, arithmetically symbolized as

2 + 2 + 2 + . . . until he counts the last two

Three-by-three, arithmetically symbolized as

3 + 3 + 3 + . . . until he counts the last three

Or (this is important), he may do it by any of several other combinations, which can be arithmetically symbolized, for instance as 6 + 3 + 4 + 2 + 5 + 4. Whatever the particular method he may use, he will count twenty-four as the new number, and this quality, which has been "quantified," is symbolized by the new numeral "24." This last operation is that of addition, one of the four basic arithmetic processes. We may now ask, Within those twenty-four red roses obtained from two bundles, can he perform the operation of subtraction? The answer is obviously yes. Similarly, we may be assured that, limited to those twenty-four red roses, he may perform the operations of multiplication and division as well.

It is now worthwhile to remind ourselves that the "twenty-four" the person counted as a total number is a quality, with this difference: it is a quality that is measurable. We may now summarize that the measurable qualities are those that are amenable to counting and can be subjected to the arithmetic operations. Such ones are quantifiable qualities, commonly expressed as *quantities*. Those that cannot be so quantified are numerous and all around us. There are several qualities, which though tracked by human perception, cannot be subject to measurement. The stink of a rotten fish and the sweetness of a lily are distinguished by the same perception. When there are more lilies gathered together, there may even result a higher intensity of the sweet smell, but it is not measurable, at least not yet. The jarring noises of a pneumatic hammer in action as well as the song of a cuckoo are both distinguished by the same perception. But there is no scale of gradation for the "sweetness" of the sound between these extremes. Similar extremes can be found in perceptions of taste and touch, wherein, despite the ability to distinguish and even give a qualitative description that "this is sweeter than that" or "this is gentler to the touch than that," there are yet no gradations or graduations conforming to unified scales of measurement. When we go beyond our perceptions, measurement of qualities becomes even impossible. How can beauty, kindness, honesty, courage, and the like be measured? Such lapses are often quoted by those bent upon pointing out the limitations of science.

3.5 Quantities and Measurements

We have noted that in the process of answering the question, How many? (when relevant), we need counting, and that counting, besides being a measurement by itself, is the basis of all other measurements. When we measure the height of a person who is, let us say, $5'10''$, we are in effect counting the number of one-inch-long pieces, placed end to end, that cover the entire length of the person while he is standing erect, from the bottom of his foot to the top of his head, which comes to be seventy in number. The height scale against which he was made to stand, and on which his height was "read," simply served the purpose of a reference, conforming to the first rule of counting, namely,

If $A = C$ and $B = C$, then

$A = B$

where A is the height of the person in inches, B is the number of one-inch-long pieces required for the purpose, and C is the height read on the scale. (Instead of having individual one-inch pieces separately in a bunch, we have such lengths imprinted on the height scale for convenience).

Now, let us say that a person's weight needs to be found and that we have a conventional weighing balance for the purpose. Placing the person on one pan, the required number of one-pound weights are placed on the other pan until neither of the two pans sinks. Then, the weight of the person is obtained by counting the number of one-pound weights required for the purpose. If instead of all one-pound weights, we used one hundred-pound weight, one fifty-pound weight, two ten-pound weights, and three one-pound weights, we do the simple addition—(1 × 100) + (1 × 50) + (2 × 10) + (3 × 1)—and obtain the weight of the person as the sum: 173 pounds.

Another "property" of the person can be so measured: the person's age, which is the number of years elapsed since the person was born. Here the counting is even more direct. Suppose on the first birthday of a person a marble were placed in a pouch, and at the end of another year, another marble were placed into the same pouch, and another marble after another year, and so on, until now. If we now want to know the age of the person, we need only open the pouch and count the number of marbles in it. The same result is obtained if we do the arithmetic of subtracting the year of birth from the current year (which indeed is the way age is most often found). Age, counted in years, is the measure of elapsed time. If the time elapsed for an event requires less time, say the timing of a 100-meter sprinter, we count in seconds and decimals of seconds with an appropriate stopwatch. The above three properties of the person, height (in terms of length), weight (mass, to be more precise), and age (in terms of time), being measurable, are "quantified." These also happen to be three of the most fundamental qualities directly measurable in the physical world. It is necessary to note that the inch, pound, and year (or second) are arbitrarily defined units. We may count the length, the weight, and the time equally well in any other unit. When so done, the numbers arrived at will be different, but with no change in the rule (or principle) of measurement.

3.6 Derived Quantities

Using only the three quantities mentioned above, we can make a lot of measurements in the physical world. For instance, confining ourselves to length alone, we can measure the area of a given (flat) surface. Consider an area whose length is 3 inches and the width is 2 inches. If we draw the lines at the markings of inch lengths, we get six area "pieces," each piece having 1-inch length and 1-inch width.

If there is another area whose length is 5 inches and the width is 6 inches, drawing the lines as mentioned above, we get thirty area pieces, each 1 inch long and 1 inch wide. Similar "trials" with areas of different lengths and different widths should have led the first observer(s) long ago to the conclusion that area can be obtained by multiplying length by width, when both are measured in the same units. The logical process that led to that conclusion we now call *induction*, and we expect the elementary school student to be capable of it. It is interesting to note that the observation above is the basis of such induction, and in that sense, this "knowledge" is "experimental" in nature. Further, we should note that the width is measured and counted in inches, in the same way that the length is measured. Also, the resulting pieces of area are measured in inches, both the length and width of which are just 1, and designated as "1 square inch" or "1 inch square." A similar construction, counting, and induction has led to the observation that volume, counted as the number of "1 cubic inch," is given by multiplying length × width × height, all in inches. The fact that using any other unit of length instead of inches results in the corresponding square and cube is a relatively minor detail. Area and volume are quantities derived from one basic quantity: length.

Even more obvious are quantities that are derived from more than one basic quantity. For example, we can distinguish between an automobile moving fast and another moving slow. But when we want to specify what is "fast" and what is "slow," that is, when we want to quantify the quality of "fastness," we need to measure speed. Unlike length or weight, speed involves two basic quantities that need to be separately measured: (1) the distance moved, and (2) the corresponding time elapsed. Then, by dividing the number of units of distance by the number of units of time, we obtain the defined quantity known as speed. Speed is thus a quantity, easily obtainable by a simple derivation, but not measurable independently; it is expressed as distance moved in unit time. Similarly, we distinguish that steel is heavier than aluminum. But when we want to specify how much heavier, we need to know the density of each, which is obtained by measuring the weight and volume of each, then dividing the weight of each by the corresponding volume.

3.7 Units for Measurement

A *unit* is a quantity used as a standard, in terms of which other quantities of the same kind can be measured or expressed. We discussed earlier measuring the height of a person and used the inch as the unit of length. The inch itself is an arbitrarily defined quantity, meaning there is nothing in nature to recommend this particular length over any other length. If we want to measure the length of a tennis court, we may use the foot as a better unit, and if we want to express the distance between Boston and New York, we may use the mile as a convenient unit. The relations among the inch, foot, and mile, meaning for instance, how many inches make up a foot, are again arbitrary decisions. These are neither nature dictated nor made inevitable by reason; they are purely manmade. But all those who use these units are bound by a common agreement, referred to as a *standard*. A standard inch is the same length whoever uses it, whenever, wherever. Similar considerations apply to units of weight: the pound, ounce, ton, and so forth. The units of time—the year, day, hour, minute, and second—are no different in this regard, except that a day is about the same duration as the earth takes to make a rotation on its own axis.

3.8 Fundamental Quantities and Dimensions

We have mentioned three quantities—length, mass, and time—with some familiar units to express these. We may at this point note that the inch, the pound, and the second used in this discussion are simply circumstantial, in that, being located in the United States, I have used the units familiar to the general public in this country, almost all other countries have switched to the metric system, which is also the basis for the International System of Units (SI) more commonly followed by the scientific community. These three quantities are so fundamental that all quantitative relations in the physical world dealing with motion and the tendencies of objects to motion—the aspect of physics known as "mechanics"—can be adequately analyzed in terms of these. To deal with all other quantitative aspects of the physical world, only four additional quantities are required: (1) temperature difference (in degrees centigrade or Fahrenheit), (2) intensity of light source (in candles), (3) electric charge (in coulombs), and (4) amount of substance (in moles). As mentioned earlier, some of the other quantities that are thus far nonmeasurable may, in the future, be made measurable, hence quantified. Besides these seven fundamental quantities, a considerable number of derived quantities are in use; we have cited speed and density.

The statement of the magnitude of a physical quantity consists of two parts: a number and a unit. The height of a person can be 6 foot, not just 6. Further, it is "foot," not "feet," because the height is six times 1 foot, or 6 × 1 ft. If we so choose, we can also state the height in inches: 6 ft. = 6 × (1 ft.) = 6 × (1 ft × 12 in.) = 72 in. Now let us consider a derived quantity: speed. An automobile requiring 4 hours to cover a distance of 240 miles obtains an average speed: length/time = 240 mi./4 hr. = 60 mph. Here we divided one number by another to obtain sixty. Further, we performed division between the two units to obtain the newly derived unit: mph. As in the case of fundamental units dealt with before, here again, a number combined with a unit expresses the magnitude of the physical quantity, speed. The magnitude, 60 mph, if we so choose, can be expressed in other units by using the required simple computation. While dealing with physical quantities, whether fundamental or derived, the units should be included throughout the computation. One may cancel, multiply, or divide units, as if they were numbers. Suppose we want to express the speed in feet per second. Knowing that 1 mi. = 5,280 ft. and 1 hr. = 3,600 sec., we proceed as follows:

1 mph = 1 mi. ÷ 1 hr. = [1 mi. × (5,280 ft./1 mi.)] ÷ [1 hr. × (3,600 sec./1 hr.)]

= 5280 ft./3,600 sec.

= 1.47 ft./s

Using this new unit, the average speed, 60 mph, may now be written as

Average speed = 60 × (1 mph)

= 60 × (1.47 ft./s)

= 88.2 ft./s

3.9 Dimensional Analysis

We have witnessed above that speed can be expressed in different units; the same speed can be expressed, if we wish, in miles per second, meters per day, inches per year, or in any other units we choose. In each case, the number will be different, combined with the corresponding unit. But there is something common among all these quantities: they are all obtained by dividing a quantity of *length* by a quantity of *time*. Symbolically we may represent this fact as

Speed = L/T

where L is length and T is time. Relative to such representation, it is said that "L/T" or LT^{-1} is the "dimension" of speed. The dimensions of some of the other simple quantities are shown below:

- Area = length × length = $L \times L = L^2$
- Volume = area × length = $L^2 \times L = L^3$
- Density = mass ÷ volume = $M \div L^3 = ML^{-3}$
- Specific gravity = density ÷ density = $ML^{-3} \div ML^{-3} = 1$, meaning, it is "dimensionless"
- Acceleration = speed ÷ time = $[L \div T] \div T = L \div T^2 = LT^{-2}$
- Force = mass × acceleration = $M \times LT^{-2} = MLT^{-2}$

A close study of the dimensions of the quantities shown above reveals some significant features:

1. Dimensions are free of units. Whatever the units used to express the quantity, the dimensions of that quantity remain unaltered.

2. Dimensions of a derived physical property are obtained by representing with symbols (with required exponents) the various fundamental physical properties conforming to the definition of the derived property.

3. Dimensions have no numerical significance. Stripped of numbers, they reveal the structure, rather than the value, of the quantity. High speed or low speed, for instance, have the same dimensions (L/T, or LT^{-1}).

4. Dimensions of the product of two kinds of quantities are the same as the product of the dimensions of those quantities.

5. Dimensions of the ratio of two kinds of quantities are the same as the ratio of the dimensions of those quantities.

6. The ratio of two quantities having the same units is "dimensionless"; it is a pure number. Both the circumference and diameter of a circle, for instance, are units of length; their ratio is only a number (π); they have no dimensions. Similarly, the specific gravity of a substance, which is a ratio of its density and the density of water, each having the same units, is a pure number; it has no dimensions.

When quantitative relations between or among parameters are attempted, it is often necessary to state such relations in the form of equations. The most significant application of dimensional analysis is to check that such equations are "balanced," as explained following.

Firstly, a comparison of two quantities is meaningful only when their dimensions are the same; 5 centimeters can be compared to 7 miles because both have the same dimensions, (L). But comparing 5 grams to 7 miles makes no sense. Similarly, we can add two quantities when their dimensions are the same. We can add 3 milligrams to 45 tons, but adding 4 inches to 6 grams is nonsense.

In the course of formulating the relation among four quantities, let us suppose that we form an equation:

$$S = X(Y^2 + Z) \tag{3.1}$$

To see that the quantities of the same units may be added in this equation, it is necessary that Y^2 and Z have the same units. And to see that the quantities of the same units are compared, $X(Y^2 + Z)$ should have the same units as S. Both these conditions depend on what the actual units of S, X, Y, and Z are. To make this issue concrete, let us suppose that the quantities have the following units:

S: mile

X: mph

Y: hour

Z: hour^2

Substituting only the units—ignoring the numerical values for the moment—we have

$$\text{Mile} = (\text{mile} \div \text{hour}) \times (\text{hour}^2 + \text{hour}^2) \tag{3.2}$$

The first condition, namely, that Y^2 and Z of (3.1) have the same units, is fulfilled above.

Then, on further simplification of (3.2), we have

$$\text{Mile} = \text{mile} \times \text{hour} \tag{3.3}$$

which is obviously wrong. If X had units mile/hour^2 instead, and all other quantities had the same units as before, the reader might find that the equation could have been balanced.

Such balancing would have been much easier if we used dimensions instead of units, for then the structure of the equation, free of quantities, would be revealed. This is done below, incorporating the changed units of *X*.

$S: L$

$X: L/T^2 = LT^{-2}$

$Y: T$

$Z: T^2$

Now, (3.1), in terms of dimensions, is reduced to

$$L = LT^{-2} \times (T^2 + T^2) \qquad (3.4)$$

$$\mathrm{L} = \mathrm{L}$$

The reader should have noticed in (3.4) that $(T^2 + T^2)$ is taken to be T^2, not $2T^2$, if we simply follow the rules of algebra. In fact, a similar jump should also have been noticed in (3.2), wherein ($\text{hour}^2 + \text{hour}^2$) is taken to be simply hour^2. Thus, yet another feature of dimensional analysis worth adding to the list of other features given previously may be stated as follows:

Whereas in multiplication and division, alphabetical terms with exponents used in dimensional analysis may be treated simply as algebraic terms, in addition and subtraction, special treatment is necessary.

Two or more similar terms, when added (or subtracted), are subsumed into the same term.

3.10 Accuracy versus Approximation

Being accurate and being scientific are often taken to be synonymous. And similarly, being approximate is taken to be contrary to being scientific. Neither of these beliefs is necessarily true. Imagine a piece of "scientific" information in which the distance between Boston and New York is mentioned as 221.8215 miles. The questions that should arise from such information are many:

1. Where exactly are the two points known as Boston and New York?
2. What path was followed between the two points?
3. What measuring instrument and methods were used?
4. How precise was the instrument?
5. At what temperature was the measurement made, and was any correction made on that account?

And so on. All such questions are relevant because any one of the above factors, or any combination of them, is more than likely to affect the number given for distance, since it is so vulnerable at the third and fourth places to the right of the decimal point. Even a very scientifically minded person will normally be satisfied if the distance is given as 222 miles and will likely feel no impulse to ask the kind of questions mentioned above.

Quite often, the temptation to be overly accurate is common not with the fundamental or primary quantities, such as distance, but with the derived quantities. For instance, if the distance is measured as 222 miles, and the time for travel is measured as 4 hours and 15 minutes, then the average speed is given by 222 ÷ 4.25, which can be calculated to be 52.23529 . . . miles per hour, with the numbers after the decimal point continuing. At this point, all that can be seen are two numbers connected by the division sign. The tendency is to forget how the two numbers were obtained and to be carried away by the love of accuracy. The numbers resulting from such calculations, when the source of data used is unknown, unseen, or ignored, are often treated with the kind of respect reserved for deities in temples, transcending the stone, metal, wood, plaster, or other materials they are made of. It is desirable on the part of the scientist not to be carried away by such unfettered, unquestioned devotion to numbers. A TV commercial presenting a smiling, good-looking woman with the statement that 77 percent of her facial wrinkles are gone because of the use of a certain miracle lotion does not tell us how the number

seventy-seven was derived. If we ever get the primary data that yielded such a number, it is likely to be far from "scientific."

Returning to the accuracy stated for speed, it cannot be more accurate than the accuracy of the "raw" information or the primary data, namely, that of distance and that of time. The number for the distance has no numerals after the decimal point and that for time has two such numerals. It is quite adequate, and in fact proper, to round off the number for speed to only two places after the decimal point. Thus, giving speed as 52.23 miles per hour is the right thing to do, and for all practical purposes, to mention it as approximately 52 miles per hour is not inconsistent with being "scientific." To compute and mention the speed as 52.23529 . . . miles per hour instead is not only a waste of time, but it is misleading. The saying, "The road to hell is paved with good intentions," may in such cases acquire some relevance.

3.11 Bibliography

Goldstein, M., and I. Goldstein. *How We Know . . .* New York: Plenum Press, 1978.

———. *The Experience of Science*. New York: Plenum Press, 1984.

Murray, R. L., and G. C. Cobb. *Physics: Concepts and Consequences*. Upper Saddle River, NJ: Prentice Hall, 1970.

Pap, A. *An Introduction to the Philosophy of Science*. New York: Free Press of Glencoe, 1962.

Richardson, M. *Fundamentals of Mathematics*. New York: Macmillan Co., 1941.

4

The Purpose and Principles Involved in Experimenting

By no amount of reasoning can we altogether eliminate all contingency from our world. Moreover, pure speculation alone will not enable us to get a determinate picture of the existing world. We must eliminate some of the conflicting possibilities, and this can be brought about only by experiment and observation.

—Morris R. Cohen

This chapter focuses on the principles that follow the basic assumption that the changes that occur in a given, confined part of nature can be explained in terms of cause-effect relations. We first note the objections raised against such assumption by philosophers David Hume (1711–1776) and Bertrand Russell (1872–1970). Then, we summarize a methodical analysis of cause-effect relations as presented by logician J. S. Mill. Following this, by way of planning for cause-effect experiments, we discuss such needs as (1) standardizing the materials used in the experiment, (2) striving for reproducibility of the experimental outcome, and (3) limiting the number of experimental runs.

4.1 The Purpose of Experimenting

The purpose of any experiment is to collect data, which, in turn, needs to be analyzed or processed to derive inferences. There are four ways of collecting data: observation, surveys, computer simulation, and experiment. Though analyses of data using these methods share common features, our interest here is confined to the collection of data from experiments. An astronomical observation, for example, is not an experiment, though it entails data collection. This is so because there is little we can do to control the independent variables, in this case, the arrangement of astronomical objects. A laboratory experiment, which is our concern in this book, is understood to be such only when the one or more independent variables, which act upon the dependent variable, can be controlled to the required levels.

Through the medium of the experimental setup, we alter one or more independent variables within the possible, or required, range, and we observe, measure if need be, and record the outcome, which either is itself or leads to the determination of the dependent variable. In a situation like this, if the independent variable is A and the dependent variable is B, it is customary to say that "A caused B," where A is referred to as the *cause* and B as the *effect.* The change made in quality or quantity (of a measurable property) on the *cause,* and the corresponding change in quality or quantity (of a measurable property) in *effect*, observed or measured and recorded, constitutes *data.* In most experiments in the physical sciences (including technology), to connect one independent variable, the *cause,* with one dependent variable, the *effect,* is fairly common. Can there be situations wherein causes A_1, A_2, A_3 . . . can all participate together to produce an effect B? The answer is yes, and this indeed is, more often than not, the case in biological, as well as some industrial, experiments. In this chapter, we restrict the discussion to *one cause–one effect* relations, but only after noting that although convenient and often adequate, these may fall short of correctness. Discussion of the combined effects of many causes will be dealt with in Chapters 7, 8, and 9.

4.2 Cause and Effect

Debate on the relation between cause and effect, *causality,* as it is known in philosophy, is more than two hundred years old and was prominently projected by Hume, a British philosopher. Lightning and thunder follow a sequence, lightning before thunder. In normal commonsense language, we may say that a particular lightning "caused" the subsequent thunder. Hume's argument against such a statement runs somewhat like this: Our knowledge of the external world is derived only through experience, which in turn is based on sense perceptions. (This brand of philosophy is known as *empiricism*). Lightning is a sense perception of the eye; thunder is another perception of the ear. In between, we did not perceive anything like "cause" because it simply did not exist. Further, thunder following lightning can be attributed to sound traveling more slowly than light. Therefore, our idea of sequence—lightning first, then thunder—is not quite correct either. Our saying that lightning caused thunder is thus only a way of speaking, a custom, nothing that "necessarily" follows from thought. Mathematical relations that "necessarily" follow from thought, free of perceptions, and, equally, experiments that can be directly conducted without relying on sense perceptions do not need any (fictitious?) "cause," hence, are philosophically supportable.

Since Hume, a lot of exposition on, as well as debate for and against, causality abounds in philosophy. The objection against "cause" was later renewed by Russell,[1] with the determination to "extrude it [the word 'cause'] from philosophical vocabulary." Russell's main objection seems to center on the time element between cause and effect, and he seems to object less if the cause-effect relation is understood to be only probable, and even when the probability is very high, Russell argues, it is not a "necessary" relation. He gives an example: "Striking a match will be the cause of its igniting in spite of the fact that some matches are damp and fail to ignite."

In the context of experimental research, however, the notion—even if it is only a notion—of a *cause-effect* relation has survived to this day, leaving aside the raging battle in the world of philosophy. In experimental research, wherein confirmation, prediction, and control within certain domains of nature are the purposes, the idea of cause is not only useful but necessary to analyze and express the events in ordinary, commonsense language. Presence or absence of current in the connecting chord makes the difference between the fan working as expected and not working. Current as an agent of action is the cause necessary to expect the effect, the functioning of the fan. Using the concept of cause between or among events is common in experimental research. The invariant relation between two events *A* and *B*, meaning that when *A* happens, *B* happens also, and when *A* does not happen, *B* does not happen either, is sufficient to say that *A* and *B* are *causally related*. If *A* always happens before *B*, there is a temporal relation as well. In such circumstances, we say that *A* is the cause of *B*, or simply, *A* causes *B*. In some circumstances, the temporal relation may not be necessary for the causal connection. The falling stone acquires acceleration because of earth's gravity. But whether gravity happened before acceleration is a question leading to unnecessary debate. This idea, from time to time, has raised its head as an objection against "action at a distance," as implied by the Theory of Gravitation. Suffice it to say that gravity is the cause of acceleration, or gravity causes acceleration.

4.3 Pertinence and Forms of Cause

A bundle of isolated facts is not science any more than a heap of bricks is the wall of a house. An "ordering" of the bricks, placing them side by side, some on top of others, is essential for building a wall. Likewise, science calls for an ordering among events; that order is mostly causal in nature. When you go to the mechanic to fix a flat tire, he looks for the nail as the first step. The flat tire, the effect, is caused by the presence of a nail in its wall. Looking for cause is so ingrained in human nature that scientists, who happen to be human, cannot do otherwise.

The relation between cause and effect has variations. Firstly, we may distinguish between a *necessary cause* and a *sufficient cause.* Exposure to tubercle bacillus is necessary for someone to get infected with tuberculosis (TB), but it is not sufficient; all nurses and doctors in the TB sanitarium are not TB patients. Lack of immunity is another cause required for the infection to become active. These two necessary causes together act as sufficient for the effect. Avoidance of either one will prevent the effect. In contrast, overeating may be sufficient cause for gaining weight.

Secondly, we may distinguish a *proximate* (or immediate) *cause* from a *remote cause.* If *A* causes *B*, and *B* causes *C*, then we think of *B* as the proximate cause and *A* as the remote cause of *C*, the effect. If I slept late, missed the bus, and, thus, could not go to the theater as I planned yesterday, sleeping late is the remote cause, and missing the bus is the proximate cause. In searching for causal connection in an experimental investigation, one or the other of the above forms of cause is discernable.

4.4 Mill's Methods of Experimental Inquiry

Occasions in which only two events, isolated from all other events, are causally connected with each other are rare. More often than not, two events mixed among many need to be detected as causally connected, and others need to be ignored as insignificant, if not irrelevant. Francis Bacon (1561–1626) is credited with an early attempt to formulate in simple rules the way the mind functions in such detectivelike exercise. After enrichment by several other philosophers and scientists, including Isaac Newton, the refined, final form of these rules, totaling five in number, is known today as *Mill's Methods of Experimental Inquiry*,[2] after the British philosopher J. S. Mill (1806–1873). These are summarized below.

4.4.1 Method of Agreement

Mill writes, "If two or more instances of the phenomenon under investigation have only one circumstance in common, the circumstance in which alone all the instances agree is the cause (or effect) of the given phenomenon."

Mill's statement being somewhat abstract, it needs some elaboration. Suppose a group of seven students went to a weekend (free!) beer-tasting party, and four of them became sick after drinking. Let us call the four students *A*, *B*, *C*, and *D* and the brand names of beer available *l*, *m*, *n*, *p*, *r*. The next day, when recovered, the students got together and pooled their information (see Table 4.1).

Table 4.1 *Students' Pooled Beer-Drinking Data*

Student Who Got Sick	Drank Beer of Brand . . .				
A	l	m		p	r
B		m	n		r
C	l	m		p	
D	l	m	n		r

In the above example, the two events causally connected are drinking beer *m* and getting sick. It would be reasonable if they found brand *m* to be the culprit and "hated" that beer and, from that day on, lost no chance to say so.

4.4.2 Method of Difference

Mill writes, "If an instance in which the phenomenon under investigation occurs and an instance in which it does not occur, have every circumstance in common save one, that one occurring only in the former, the circumstance in which alone the two instances differ, is the effect, or the cause, or an inseparable part of the cause, of the phenomenon."

Extending the Method of Agreement, suppose subsequent to the four students blaming beer *m,* student *A* met another student, say *E*, who went to the party that night but did not get sick. On being asked by student *A*, *E* told *A* that he drank beers *l*, *p*, and *r*, but not *m*. "No wonder you did not suffer like us. That accursed beer *m* should be banished," is what *A* could say (see Table 4.2).

Table 4.2 *Students A and E Find that Beer m made the difference.*

Student	Drank Beer of Brand				
A	*l*	m		p	r
E	*l*			p	r

The event that made the difference was the drinking of beer *m* in one case and not drinking it in another. And the causal connection is between the above difference and the event of getting sick. Drink beer *m*, and get sick; don't drink *m*, and don't get sick

4.4.3 Joint Methods of Agreement and Difference

The evidence from the above two methods constitutes the present one. It may be formulated as

$$HJK\text{—}t\,u\,v \qquad\qquad HJK\text{—}t\,u\,v$$

and

$$HRS\text{—}t\,w\,x \qquad\qquad JK\text{—}u\,v$$

The above two pieces of evidence can be summarized as

- When H is yes, t is yes.
- When H is no, t is no.

The causal connection, obviously, is between H and t (t is an "inseparable" part of H). Though in the above demonstration we have chosen three terms ($H\ J\ K$, $t\ u\ v$) as the greatest number of terms, and two terms ($J\ K$, $u\ v$) as the least, in principle, these numbers can vary, and the method will still apply.

Example:

HJ— tu		HJ— tu		HK— tv
	and		and	
HK— tv		J— u		K— v

In this example, two terms and one term are used to establish the causal connection between H and t, though by a slightly different path.

4.4.4 Method of Residue

We may often find a cause as one of many bundled together. Likewise, we may also find the corresponding effect as one of many bundled together. Each of the bundles can be likened to the ends of electrical cables, consisting of many wires, the cables to be connected together. If all but one wire from the first cable are connected to all but one wire from the second cable, that the remaining wires from each are to be mutually connected is logical. That is the kind of logic involved in the Method of Residue.

Considering small numbers for convenience, although the method remains the same, an electrician is required to connect two cables, each consisting of three wires, one black, one green, and one white. He does his job as a routine, connecting black to black, green to green, and white to white. Now, suppose the first cable consists of black, green, and white wires, but the second cable consists of blue, red, and yellow wires. The job is now not so obvious. A little pause and some inquiry are necessary; according to his local "code," the black wire is hot, the white wire is neutral, and the green wire is grounded. If he finds on inquiry that, by the code of the foreign country from which the second cable was imported, blue is neutral and yellow is ground, then he connects the white wire of the first cable with the blue wire of the second and the green of the first with the yellow of the second. Yet to be connected are the hot black wire in the first cable and the hot red wire in the second cable. This last connection is the connection between *residues,* which is now obvious to the electrician. Likewise, in terms of logic, we can identify which particular cause in the bundle of causes is responsible for a particular effect in the bundle of effects; this is done by relating the *residue* in the bundle of causes with the *residue* in the bundle of effects.

Mill writes, "Subduct from any phenomenon such part as is known by previous induction to be the effect of certain antecedents, and the residue of the phenomenon is the effect of the remaining antecedents."

This may be symbolized as below:

$L\,M\,N$—$x\,y\,z$

L is known to be the cause of *x*. *M* is known to be the cause of *y*. Thus, *N* is the cause of *z*.

It may be noted that, unlike the three previous methods, wherein *causal connection* was self-sufficient, in this method, we need to mention *cause* and *effect* separately, thereby implying temporal relation. This is because the Method of Residue explicitly appeals to the previously established pair(s) of cause and effect. Also, note the assumption that, although the causes and effects occur in separate bundles, for a given effect, only one of the causes is responsible for the effect.

4.4.5 Method of Concomitant Variation

This method is similar to the Method of Residue in that a bundle of factors is causally related to another bundle of factors. Unlike the Method of Residue, however, we do not know beforehand the cause-and-effect relation between all but one pair of the wires in the first bundle and those in the second bundle. In this method, though only one of the members in the bundle of likely causes is responsible for a given effect under observation, we do not know which member that is. Further, the problem can be compounded in that it is impossible to eliminate completely the suspected cause from the bundle of likely causes. Under circumstances in which this method is relevant, we can impose, or wait for nature to bring about, a quantitative variation over a suitable range in the suspected cause and observe whether this results in a quantitative change in the effect under observation. If such a change is factual, the inference is that the suspected cause is, indeed, the cause of the effect under observation.

An extract from Mill's description of this method reads, "Though we cannot exclude an antecedent altogether, we may be able to produce, or nature may produce for us, some modification in it. By a modification is here meant, a change in it not amounting to its total removal. . . . Whatever phenomenon varies in any manner, whenever another phenomenon varies in some particular manner, is either a cause or an effect of that phenomenon."

Using a plus sign (+) for an increase and a minus sign (–) for a decrease in the degree or quantity of the individual members of the pair of factors in question (described earlier), and noting that an increase in the first member of the pair may result in either an increase (*directly* related) or in a decrease (*inversely* related) in the other member of the pair, the following two relations can be symbolically formulated:

1. Directly related

   $A\ B\ C—p\ q\ r$

   $A\ B\ C+—p\ q\ r+$

   $A\ B\ C-—p\ q\ r-$

   Then, C and r are causally connected.

2. Inversely related

   $A\ B\ C—p\ q\ r$

   $A\ B\ C+—p\ q\ r-$

   $A\ B\ C-—p\ q\ r+$

Then (also), *C* and *r* are causally connected.

The Method of Concomitant Variation, it should be noted, is the only one among Mill's five methods that specifies the degree or quantity of a phenomenon, hence, requires quantitative experiments in which the phenomenon needs to be varied (or observed) over a range. This principle is extensively used in what may be called experiments of parametric control for gathering empirical data. The other four methods have an "all or nothing" character, hence, are qualitative in nature.

4.5 Planning for the Experiment

Experiments are often time-consuming, if not also expensive. It is reasonable, then, to plan the goal, extent, and step-by-step course of experiments before embarking on their execution. Since science is not a product of a manufacturing process, the unforeseen and the unavoidable are to be expected. But the element of contingency should be reduced to the minimum possible. An unplanned, haphazard beginning can be rather wasteful, both in time and money, besides possibly leading to a dead-end. In preparing to conduct experiments, one needs to attend to several details, some obvious, some not so obvious.

A large number of experiments consist in correlating an independent variable, a cause, with a dependent variable, an effect. Between the cause and the effect, speaking in terms of strategy, is located a device, usually called the *experimental setup*, whose function is to receive the input in the form of cause and to produce the output in the form of effect. The input and output, either one or both, may involve some materials or objects that need to be chosen and may not be arbitrary. The required means is *sampling*. It may also happen that the input and output, either one or both, may not be tangible objects but measurable quantities, such as pressure, temperature, and voltage. In either of these cases, planning the experiment is relatively straightforward. On the other hand, in such areas of study as agriculture, patient care, and industrial quality assurance, in which several variables, or causes, simultaneously influence the desired output(s), or effect, the planning of experiments can become complex and confusing. While a considerable amount of literature has grown up around the various aspects of preparing to conduct experiments, an analysis now known as "Design of Experiments" is indispensable for planning such many-cause experiments as mentioned above. The earliest attempts at design seem to have originated with the scientific study of agriculture; the guesswork of Bacon relative to crop yields from seeds subjected to different treatments (one of those was to soak seeds in urine) is an example. Systematic work based on sound principles of experimentation, again in agriculture, began in the early part of the twentieth century. Some of the basics involving more than one cause are dealt with in Chapters 7, 8, and 9. The few, basic principles of planning for experiments discussed in this chapter are somewhat general in nature and can be applied with benefit to experimental work in many diverse areas of study.

4.6 Standardization of Test Material(s)

To the extent that the experimenter is specific about the materials involved, his experiments are likely to be reproducible. The degree of involvement may vary from the highest-extreme to the lowest: from the materials being tested as a part of the experiment to the materials used for the furniture in the room. To illustrate the point, we restrict our discussion here to the "closer" materials. (There are records of the "farther" materials having significance—discovered too late—rendering the experiment worthless). Let us take a common material: water. Suppose part of a hypothesis is to use water to make a certain mix, which provides the sample for specimen preparation; for example, it says to add 30 cc of water and mix thoroughly. The hypothesis is unspecific in that it uses the word "water" in a generic sense. Could it mean water from the faucet, even if it looks muddy? Could it mean freezing cold water? What if the water is 200°F? Instead, if we find in that part of the hypothesis, add 30 cc of distilled water at room temperature, to that extent, the water used in the experiment is "standardized."

One can, on reflection, recognize that many materials known to us and often involved in experiments are known in their generic form. Clay, salt, steel, plastic, and wood are some examples. For instance, we may find mention of "concentrated hydrochloric acid," without knowing the degree of concentration. When such generic materials are involved in the experiment, the investigator should make sure to specify the materials in terms of their significant, determinable properties and either stock enough of the material and some to spare for all the experiments involved (assuming that the material does not decompose or deteriorate), or check every new consignment to be sure that the measurable variation in the specific properties are within predetermined limits.

4.7 Reproducibility

The purpose of any experiment is observation. But it is not a casual observation, loosely related or inconsequential. It is an observation made to use as evidence in the process of building up a case, that case being the hypothesis concerning the connectedness between selected events of nature. As a witness in a court case is expected to give the same answer to the same question asked by the trial attorney, time after time, so the experimenter expects, from that part of nature that is "fenced in" by his experimental setup, to yield the same response to the same stimulus created in his experimental condition. The experimenter acts in a certain way in terms of his experimental variables, and he expects nature to react the same way, time after time, whatever that reaction may be. In addition, if the experimenter creates the same condition elsewhere and repeats his actions, he expects the reaction from nature to be the same as in the previous site of his experiment. Thus, the observation made by the experimenter is expected to be independent of both time and place.

There is nothing to assure the experimenter that this requirement "should" happen based on some inevitable, more fundamental principle, or law of nature. That the sun has come up every morning all these years is no guarantee that it "should" happen tomorrow also. Instead, it is an act of faith, with which we all function. The experimenter, likewise, needs to proceed with this faith in *repeatability*, for without such faith, no experimental research is possible. The philosopher Mill termed this, without antecedent, the Law of Uniformity of Nature.

Somewhat connected, but in a different respect, is what is known as the *reproducibility* of experiments. Everything that is published is not holy. Suppose that after some time and in another place, another experimenter wants to make sure that under reasonably similar circumstances as those published, his actions in terms of variables and the reactions of nature toward those actions will be the same as he found reported. To the extent that this latter experimenter faces no surprises, the experiment is considered reproducible. Now, in addition to time and place, we may include the experimenter himself as the third element, from which the experimental observation should be independent. This stipulation is necessary for the growth of science, since, as we have said previously, science is a complex, but ordered, ever-growing structure of knowledge, involving the work of many people widely distributed in place and time. It is a self-imposed responsibility of each scientific worker to see that his work is reproducible, and only then is the work made public through whatever means. The experimenter finds himself to be a trustee.

In passing, we should note that in philosophically precise terms, nothing is reproducible in nature since every event is unique, and nothing ever repeats. "The moving finger writes and, having writ, moves on." In view of this, words like "reproducibility," "repeatability," and "same," among others, are used loosely, and we tolerate, even accept, them. Instead, using words like "similar" or "identical" may be more proper.

4.8 Number of "Experiments"

As pointed out above, absolute reproducibility is perhaps impossible. But reproducibility to a satisfactory level, as a matter of faith, is necessary. As to how many times a given relation between a cause and its effect needs to be confirmed to be accepted as reproducible, this is a question that should not be pressed too far. On the other hand, experiments whose value rests on reproducibility may be called "experiments with parametric variation." These are experiments conducted with the goal of establishing a relation, often algebraic, between a cause and its effect. Such relations, often in graphical form, abound in the literature of science and engineering. Conventionally, in such graphs, the horizontal (x-axis) shows the independent variable, the cause, and the vertical (y-axis), the effect. The number of experiments required to plot one such graph depends on the nature of the relation between x and y. If only a few experiments, within appropriate levels of x, indicate a straight-line relation, the experimenter is emboldened to restrict the experiments to a small number. If, on the other hand, the relation between x and y shows a nonlinear trend, more experiments, at closer intervals of x values, will be required. Further, if either a nongeometric curve or a combination of curves is the trend, the number of experiments needed will be even more. As to exactly how many experiments will be required in each case, there is no rule. All that can be said is, the greater the number of experiments, the greater the dependability or credibility of the x-y relation. A range of ten to twenty for straight-line relations, and more for others, may be mentioned, arbitrarily, as reasonable. In Chapters 16, 17, 18, through 19, we return to these concerns in greater detail.

4.9 References

1. Bertrand Russell, *Mysticism and Logic* (On the Notion of Cause) (Doubleday, Anchor Books, 1957), 174–75, Garden City, NY.

2. J. S. Mill, A System of Logic, Introduction to Logic, Irving M. Copy and Carl Cohen 8th ed. (New York, NY: Macmillan, 1990) 377–411.

4.10 Bibliography

Cohen, Morris R., and Ernest Nagel. *An Introduction to Logic and Scientific Method.* Boston: Routledge and Kegan Paul, London: 1961.

Johnson, W. E. *The Logical Foundations of Science.* Mineola, NY: Dover Publications, 1964.

Wilson, E. Bright, Jr. *An Introduction to Scientific Research.* Mineola, NY: Dover Publications, 1990.

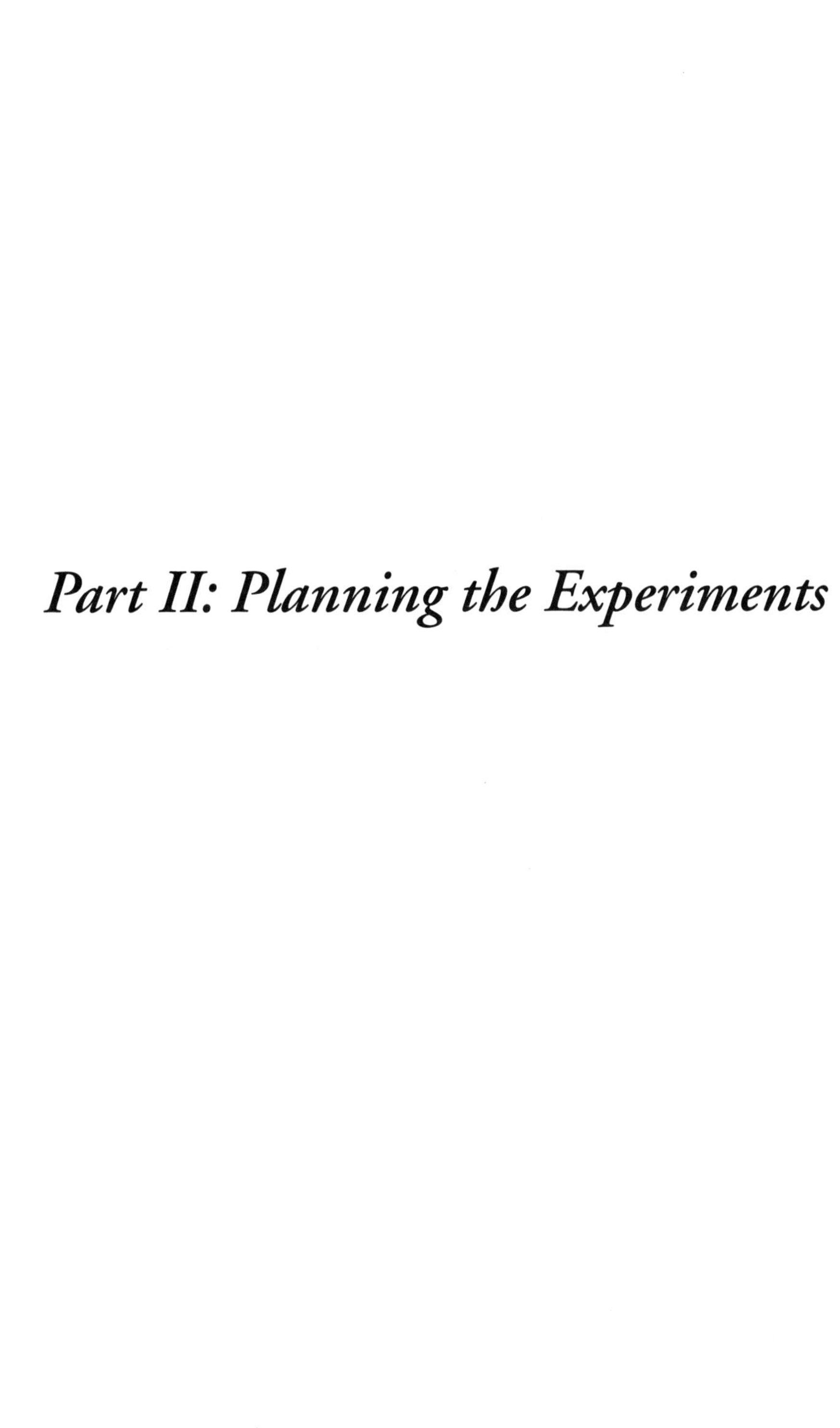

Part II: Planning the Experiments

5

Defining the Problem for Experimental Research

Where should I start? Start from the statement of the problem. . . . What can I do? Visualize the problem as a whole as clearly and as vividly as you can. . . . What can I gain by doing so? You should understand the problem, familiarize yourself with it, impress its purpose on your mind.

—*G. Polya*

In experimental research, "problems" are questions that may be asked about the behavior of an isolated part of nature. Only those problems that demand priority and, among those, only those that can be afforded within available means, are experimented on. Money, time, equipment, and human resources are the important means. There needs to be a compatible match between the problem at hand and the person who strives for the solution. Some of the important aspects of this duality—the problem and the experimenter—are discussed in this chapter.

5.1 To Define a Problem

The word "problem," in the context of research, defies the dictionary definition, which connotes difficulty, doubt, and hindrance. While engaged in searching in new territory, that is, researching, a struggle involving difficulty, doubt, and hindrance is to be expected. But that is not what the word signifies. Instead, it serves as a loose definition to indicate a certain subject area or topic of intellectual interest with implied (but shifting) boundaries, within which an exploration of, or cause-effect relation in, nature is expected. Using the analogy of hunting, the problem is the game animal, now showing, now hiding, always tempting; the researcher is the pursuer, the hunter, the tempted. Thus, any investigation or research is necessarily linked to a problem; the nature of the research is dictated by the nature of the problem.

Confining ourselves to science, we may think of two kinds of problems: those in pure (or fundamental) science, and those in applied science. Problems in pure science are aimed at getting a better understanding of certain segments of nature than we so far have, nature being meant to include all aspects of the animate and inanimate world, including humans. Fragmented into many aspects, and each aspect having only ad hoc answers to several questions we may ask, there is ample scope for generating problems to engage several researchers. Almost always, more attempts to answer lead to more questions. Some answers may be obtained by careful observation of the particular aspect of nature as it operates, unaided or undisturbed to the extent possible, by the observer. Such research tends to be *theoretical.* It may so happen that such observation can be better explained in terms of mathematical relations, a good example being the laws of planetary motion.

On the other hand, many of the problems of pure science, aimed at confirmation of hypothetical answers, as well as almost all problems of applied science and technology, are *experimental* in nature. In these, man creates a setup in which that segment of nature under examination is "fenced in" so that he can observe its behavior in reaction to several actions he can impose. Quite often, in such experiments, some kind of counting or quantitative comparison is involved, which we may call *measurement.* Most problems of applied science depend on observation, measurement, and the relation among events—qualitative, quantitative, or both. Hence, it is in such problems that experimental research gains significance. The numerous questions that we may ask about nature, most of them details in segmented domains, each constitute a problem. Some answers to questions—"solutions" to problems—fortunately, we already have. But the yet-to-be-answered questions abound, and their number keeps increasing with no end in sight. Most of the questions in applied science are "use" based and, as such, out of the, say, hundreds of questions that can be asked about a particular subject, only some are significant from the viewpoint of current priorities; other questions are either unknown or can wait their turn to become problems.

5.2 Relation of the Problem to Resources

Depending on the resources available and urgency experienced, it should be reasonable to expect that any given individual or organization will identify the problem or problems for research for the immediate future. Resources include willing and capable people, support facilities like lab space, a library, and the money to buy or build the required equipment and materials involved. Conversely, to make use of such resources, it is imperative that there should be problems. It is understandable that the only person who gets to know the intricacies and subtleties of the problem is the researcher. Hence, a problem may not have been defined in all its details before the researcher arrives on the scene. In outline, however, even as a sketch, the problem should be known to the person (or organization) who is responsible for putting the resources together, be it a guide, director, supervisor, or professor.

We may think of the problem as the meeting point of three mutually dependent driving forces: (1) the problem itself as defined earlier, (2) the problem solver, namely, the researcher, and (3) the immediate beneficiary, if and when the solution is found. It is possible that a single individual embodies all these forces, which is a very happy situation. Perhaps, research of the highest order, resulting in landmark theories in the basic sciences, is of this kind, the benefit here being mostly intellectual. On the other hand, the experiments of Thomas Edison in his early years could also approximate the same ideal, the benefit being mostly material. Unfortunately, most researchers do not fit such a description. Many graduate students take up research as a means to earn a degree as a qualification for an academic career or research positions in an organization. Depending on the funding acquired by professors, students agree to work on certain previously identified problems with a limited degree of flexibility. A graduate student entering a university or other research-oriented institution of higher learning and given free rein to research any problem he or she likes is rather rare. Some of the well-known research institutions employ hundreds of career researchers, and most of the work they do offers little room for idealism. The researchers are, more or less, assigned the problems, often within a time frame, the problems having been accepted on the basis of profitability. Industrial research and development (R&D) is another area in which researchers wear tight jackets. The products and priorities dictate the problems. It is not uncommon for a researcher to be shifted from one problem to another, often never again seeing the problem he or she made some progress with or "nearly finished." Many R&D researchers accept even that as "lucky," compared to the fairly common situation where a boss, usually a "manager," must balance the budget and casts his or her eyes on the researcher as a dispensable, nonproductive entity, preferring any other production or maintenance personnel. Getting the paycheck becomes a greater priority for the researcher than fighting for a selected problem. As research work became wider spread and more a profession than a pleasure, it became institutionalized.

Even though the present-day researcher is, more often than not, enmeshed in an organization, the need to make many decisions, particularly about the layout and details of the research problem, is very much a part of his responsibility. The problem, at best, may be handed over to him, somewhat like a sketch for a building to be built. The task of preparing a plan, equivalent to the blueprints for the builder, is left to him. To push the analogy further, whether the building is a country residence or a suburban apartment complex or a multistory office building in a metropolitan downtown area determines the considerations to be incorporated into the blueprints. Each of the above types of buildings is a specialized job. If the builder is not a big company with many specialists, but an individual with limited resources of all kinds, including experience, then he, first of all, must decide whether the job, if offered, is within his capacity to fulfill. The individual researcher is in the same position.

Often it may happen that a researcher, either out of desperation or overconfidence, will end up as the custodian of a problem, even before he or she has given any thought to its essential features, its extent, or the match between the problem and himself. A few remarks commonly applicable to all the researchers in the process of making a commitment to specific problems are discussed as follows.

5.3 Relevance of the Problem

Though researching is an experience over and above the subject matter, and the ability to separate the relevant from the irrelevant is a part of such an experience, the researcher's work will be more fruitful, and his preparation time more reasonable, if the problem is within the broad domain of the researcher's interest. Shifting from problem to problem can be frustrating. When it comes to shifting from one broad area of interest to another, the waste in terms of intellectual capacity is unjustifiable. To say no, even when circumstances compel one to do otherwise, is, in the long run, beneficial to the research as well as to the researcher. This is not meant to imply that the researcher should fence himself into an intellectual territory beyond which he should be wary of straying. What is meant is this: if the researcher is led by interest and is attracted into a related but different area, the benefits of cross-fertilization are likely to ensue; if, instead, he is pushed into the new area by sheer circumstances, motivated by financial need or gain in prestige or authority, the result, in the long run, is not desirable.

5.4 Extent of the Problem

Every problem may be visualized as a journey to be undertaken by the researcher. There is a point of departure and a point of destination. Even at the point of departure, when the researcher is likely to begin, there may be a number of uncertainties relative to the preparation necessary. The longer the journey, with the length being measured both in terms of time and money, the more thoughtful and deliberate the preparation needs to be. How long, in terms of months or years, is the problem likely to engage the researcher? The total volume of work involved, considering unforeseen snags, needs to be included. The main phases of the research activity, namely, study, equipment building (if necessary), experiments, analysis, and presentation: all of these need to be accounted for. This is not to say that the time can be calculated similarly to computing the number of hours needed by a machinist for a manufacturing job. No. Research is a totally different kind of activity; sometimes the work progresses so quickly that even the researcher is surprised, while on other occasions, days and weeks may pass in fixing small snags or waiting for small pieces of equipment or material(s). Nonetheless, the researcher should use all his experience, any available advice, and all other resources to make an honest attempt to tag a problem with a time frame. Even while doing this, he should be aware that, almost as a rule, everything takes more time to execute than estimated; hence, an overhead time needs to be added.

One of the main questions for estimating the time required asks, Is the problem a logical continuation of one or more of those previously solved? If the answer is yes, the researcher's preparatory work consists of studying and getting familiar with the previous work, published or otherwise. This saves a lot of time that would have been spent studying to define and hypothesize the problem, if it were a totally new one. If the present problem consists of extending previous work, a distinction needs to be made as to whether the extension is parallel to or in series with the previous work. If, for instance, the previous problems were to study structures by X-ray diffraction of twenty substances, and the present problem is to study the structures of ten other substances, the problem may be considered a parallel extension; the same or nearly the same equipment and technique may be required, the difference consisting mainly in certain numbers. If, instead, the present problem is an extension in series, the problem is likely to be more difficult and time-consuming. The problem may involve more thought and study, more refinement in equipment, more sophistication in technique, if the extension builds upon previous work.

The reader may wonder, knowing that research cannot be planned as a building can be planned, why researchers need to be so concerned about time. Nobody's time is unlimited. Besides, only a small percentage of research workers do not have to account for their time. In addition, those to whom the researcher is expected to report are quite often not researchers themselves; they are managers of some sort, whose expertise is confined to accounting for money and who know that time is another form of money.

The volume of work needed to solve the problem, though related to the time required, has an independent significance. A report, either at the end or at a convenient point during the course of research, may need to be presented, perhaps in the form of a conference paper, a journal article, a classified report to the sponsors, or a thesis for a degree. In the last case, for instance, no

Ph.D. candidate, however concentrated his findings may be, is expected to submit a ten-page thesis. If in contrast, the problem and its solution are to be presented in conference paper, no audience can take a forty-page presentation. The researcher may wish to live in a world of his own, in which ideals dictate that the volume of the problem and the time required to solve it are the prerogatives. Alas! the reality is that the researcher is a small part of a big world, and his independence is not unconditional.

5.5 Problem: Qualitative or Quantitative?

It is desirable for the experimenter to reflect on what kind of inputs and outputs he is likely to deal with in the proposed research problem. Is he going to deal with qualities or properties known to have been measured, or is he left to deal anew with qualities whose measurement is not well established? A little digression may be necessary to show the relation between quality and quantity, or more precisely, what we may call *qualitative* and *quantitative* research.

In the statement "Michael is tall," being tall is a quality that Michael has. The quality here is given by the adjective "tall." But the statement, being relative, lacks precision. If the Michael referred to is one of the NBA players, he could be 6′6″, or in that range. If, instead, Michael is a fourth grader in a local school, he could be only 4′4″, and still the statement is proper. The use of the word "tall" is relative, ambiguous, and nonspecific. If, instead, the statement is "Michael is 6′6″," we understand that he is taller than quite a few men who are not basketball players, but not the tallest NBA player. Here the tallness is quantified. There is no room for ambiguity. Michael's height can still be compared with that of others, but with specifics such as "three inches shorter than Kareem" or "two inches taller than John." There are many statements in routine usage where a qualitative statement, such as "George is rich," "The rose is red," or "You are late to work," is quite adequate. The degrees of George's richness or the rose's redness or your lateness are not quantified. In the last mentioned instance, whether you walked into the office a second later or an hour later than you were expected to be in, the remark that you are late to work is equally correct. Needless to say, statements with quantities are preferable in experiments, for with those, you know where you stand.

To go a step further, there are qualities that can be quantified, like degrees of tallness (the height of a person), lateness (so many minutes past an appointed hour), or hotness (the temperature of an object). What is common in all these examples is that these qualities are measurable. By contrast, in the statements "Helen is beautiful," "This fruit is delicious," or "This rose is red," we cannot measure the quality of beauty, taste, or redness. These are certainly qualities we can recognize and distinguish, but there are no procedures for, and no units of, measurement. In summary, we may say that not all qualities are measurable. In terms of measurability, we may distinguish two classes of qualities: those that can and those that cannot pass the criteria of measurability. It should be noted, however, that criteria not currently available may in the future be developed and accepted as normal, and new procedures for their measurement may come to be accepted as "standard."

It is very important for the experimenter presented with a problem to know what kinds of qualities or properties he is expected to deal with. If they are measurable qualities, his task is reduced to one of identifying the procedure and equipment needed for measurement. He must address whether such equipment is available on the market or he must devise his own, what range of variation is involved, what degree of accuracy is required, and so on. If, on the contrary, he expects to encounter qualities or properties that are not measurable by any currently known procedures with appropriate equipment, he should be candid with himself and with others, and declare that the problem is not worth proceeding with.

5.6 Can the Problem Be Reshaped?

When we talk about reshaping a problem, we are not talking of exchanging one problem for another. If this were possible without obligation, the researcher's life would be a picnic. Changing jobs where jobs are plentiful is fairly common; one is not required to explain the event. We are talking here about the possibility of the same problem changing its shape, size, and even some features as one progresses toward solving it. This process is somewhat like a hiking journey toward a mountain peak. First, the distance is a lot greater than it appears to be. Second, a lot of details not seen before spring up, as if from nowhere, with at least some of the details being unsightly. The shape of the peak, which looked smooth and rounded, perhaps one of the attractions from afar, slowly disappears, exposing edges and sharp corners. Or a certain feature, like the "Old Man of the Mountain," fascinating from a particular spot far below, when approached, may be exposed as simply a heap of rubble. And it is possible, when within touching distance, to correlate the features from afar with the details of the rubble at hand. This last aspect, in terms of intellectual understanding, is perhaps the kind of thrill that justifies for the researcher his or her struggle for a solution.

5.7 Proverbs on Problems

A fitting conclusion for this chapter may be a few passages taken from a very interesting little book (now a classic) by G. Polya,[1] which any student of science, particularly researchers, should find instructive and interesting. Among hundreds of passages one can find worthy of using as proverbs, only twelve are given below in the four stages of research problems. (Note: these "proverbs" are taken from various sections of the book, and the words are not reproduced exactly.)

General

- It would be a mistake to think that solving problems is a purely intellectual affair. Determination and emotion play an important role.
- In order to solve a problem, it is not enough to recollect isolated facts. We must combine these facts, and their combinations must be well adapted to the problem at hand.
- Do not forget that human superiority consists in going around an obstacle that cannot be overcome directly, in devising some suitable auxiliary problem when the original one appears insolvable.

At the Beginning

- Incomplete understanding of the problem, owing to lack of concentration, is perhaps the most widespread difficulty in solving problems.
- Visualize the problem as clearly and as vividly as you can. For the moment, do not concern yourself with details.
- Have you seen the problem in a slightly different form? Could you use it? Could you use its results? Could you use its method?

In Progress

- If you cannot solve the proposed problem, try to first solve some related problem.
- Consider your problem from various sides and seek contacts with your formerly acquired knowledge.
- Desiring to proceed from our initial conception to a more adequate, better-adapted conception, we try various standpoints and view the problem from different sides. We could hardly make any progress without variation of the problem.

At the End

- Did you use the whole hypothesis?
- Have you taken into account all essential notions involved in the problem?
- Looking back at the completed solution is an important and instructive phase of the work. He thinks not well that thinks not again.

5.8 References

1. G. Polya, *How to Solve It*, 2nd ed. (Princeton, NJ: Princeton University Press, 1973).

5.9 Bibliography

Cohen, M. R. *Reason and Nature: An Essay on the Meaning of Scientific Method.* Mineola, NY: Dover Publications, 1978.

Diamond, W. J. *Practical Experimental Designs for Engineers and Scientists.* New York: John Wiley and Sons, 2001.

Johnson, N. L., and F. C. Leone. *Statistics and Experimental Design in Engineering and Physical Sciences.* Vol. 1. New York: John Wiley and Sons, 1964.

Mead, R. *The Design of Experiments.* Cambridge, U.K.: Cambridge University Press, 1988.

6

Stating the Problem as a Hypothesis

> *[I]t is by mathematical formulation of its observations and measurements that a science is able to form mathematically expressed hypotheses, and it is through its hypotheses that a natural science is able to make predictions.*
>
> —*David Greenwood*

Hypotheses are such questions, relating to nature, as, Can this be true? Will this happen? Sources of such questions may be either curiosity or vested interest, and more often a combination of both. Whether originating from an individual or an organization, hypotheses are forces that impel progress in research work. The relation between hypothesis, experiment, confirmation, and theory forms the theme of this chapter.

6.1 The Place of Hypothesis in Research

In view of the fact that a hypothesis is central to any scientific investigation, theoretical or experimental, it is necessary to study hypotheses in more detail. In this chapter, we will see, among other things, the "provisional" nature of hypotheses, meaning that there is nothing permanent, nothing final, about any hypothesis. But without hypotheses, there is no meaning, no purpose, to scientific investigation. Another way of saying this is that the hypothesis, the means, impels scientific investigation, the end. Like the propeller of a ship, which is hidden from view deep down but is nonetheless absolutely necessary for the ship to move, the hypothesis is essential, subtle, and not always obvious.

Its somewhat elusive nature makes it difficult to look at close up; that close-up look is attempted here.

There is a seeming exception to the above statement, namely, investigations in the classificatory sciences, like zoology and botany. Let us imagine that a botanist on his morning walk spots a particular plant, the like of which he has never seen before. His curiosity leads him to contemplate where, in his knowledge of classification, this new plant belongs. He considers, in order, to what family, genus, and species this plant should be assigned. And once that is done, his scientific curiosity takes a vacation until he reaches his home, lab, or library. If he is an active, curious scientist—let us hope he is—he goes on to "confirm" his classification. How does he do it? There are features—let us call them "properties"—common among all members of a family. He sets himself the task of studying the particular plant (or parts thereof), to see if this particular plant has the properties that characterize a family. If it does, he then feels elated and performs similar tasks at the next levels of genus, then species. At each level, when he is doing this task, he asks himself, for example, Could this be a member of species *A*? This is a hypothesis. Then, he does the investigation to compare the properties. If the answer is yes, he accepts this as confirmation of his hypothesis, and his investigation for the present ends there. If the answer is no, he comes up with another hypothesis: This particular plant is a member of species *B*. Again, the process of investigation begins and ends with a yes or a no, with confirmation or rejection of the hypothesis at hand.

This example is given here to stress that hypotheses are involved even in simple investigations. The above investigation is mostly theoretical in that—let us assume—our botanist is using only his library resources (including his own collection of specimens); that is, he is using already-known information.

Let us imagine a similar instance of a geologist finding a piece of rock, the like of which (geologically) he has never seen before. He then may require some tests to understand the rock's physical properties and the reaction of some chemicals to it. Here again, the hypothesis is propelling him to do one test or another, but the nature of investigation here is "experimental," though not completely so. Actually, the experiments he is doing are based on his knowledge of existing information about most other minerals. In this case, the hypothesis is based on available "theory," and he devises an experiment to confirm or reject his own hypothesis.

The sequence in the case of the botanist can be summed up as theory, hypothesis, confirmation or rejection. The sequence in the case of the geologist can be summed up as theory, hypothesis, experiment, confirmation or rejection. Thus, in the case of the geologist, the hypothesis involves the theory as well as the experiment.

Both the cases mentioned above involve what we may call *preliminary hypotheses*. It is often possible that some tests will be conducted to get a "feel" for the kind of question the investigator has in mind, when he is still uncertain about the right terms in which to form the question. Based on such a feel, the investigator may form a tentative hypothesis. This may happen more often with inventors, whose investigations are intended for patent rights, than with scientists, though in most cases, inventions are meant to be "answers" to specific gaps in knowledge. Thus, again, hypothesis leads to experimentation.

Let us imagine an investigator in the early days of electricity. Suppose he is "playing" with the measurement of current in simple circuits, using a battery of known voltage across terminals of conductors. It is his purpose to discover which is the best conductor among four different metals, *A*, *B*, *C*, and *D*. But he faces a problem. His record (logbook) shows that on different days, he has used the same lengths of metal *A*, but he got different readings of current. He suspects that something is varying in those tests, which is causing the different currents. He now wants to do some *controlled* experiments. He measures the voltage every time the current is recorded; it does not vary. He double-checks the lengths of wires; they are constant within the accuracy of his measurement. He gets a clue that diameter may be the variable. He is now moving to another level of controlled testing. He keeps the voltage of the battery and the length of the wire *A* the same every time, but deliberately uses wires of different diameters and records the corresponding current readings. A tabular record shows that when thinner wires are used, the current readings are lower. At this stage he defines a new concept, *resistance*.

He is now in the situation where his guess has turned out to be correct. Reaching the next level of curiosity, he may think, I now know that smaller-diameter wires obtain less current; these offer greater resistance. But can there be a quantitative relation between the diameter of the wire and the resistance offered by the wire? Now he has reached the fourth level of his investigation relative to the resistance of wires. This time, he selects wires of type *A* in several different diameters, controls the voltage and the lengths to remain the same, and gets several pairs of values of diameter against current. He states these values in a graph. He is pleased that the points have a trend close to a smooth curve, but he is a little disappointed that the relation is not a straight line.

The experimenter now feels puzzled and reflects, Suppose I consider the areas of the cross-section of the wires instead of the diameters. He does the necessary, simple calculations, plots another graph, this time plotting areas against current. He sees a straight line! He does not run naked in the street like Archimedes shouting, "I found it," but he is very pleased with himself, as a scientist can be, though rarely.

This narration shows many iterations of experiments, from hypotheses to the final theory. To sum up the sequence:

Problem: With the same length wire *A,* why are there different currents?

(i)*Hypothesis:* There may be a variable; a closer look is necessary.

Experimental finding: With the same voltage and the same length of wire, the current is different for different diameters.

(ii)*Hypothesis:* Wire diameter is a significant variable.

Experimental finding: Thinner wires offer more resistance.

This is a *quantitative relation* in terms of a concept newly defined, resistance.

(iii)*Hypothesis:* Possibly there is a quantitative relation between wire diameter and resistance.

Experimental finding: The graphical relation between diameter and resistance is a smooth curve (good) but not a straight line (bad?).

(iv)*Hypothesis:* The relation between resistance and the area of a cross-section of wire is more direct.

(*Theory*): (No experiment here; instead a theory involving some calculation is used.)

Confirmation: Resistance is inversely proportional to the area of cross-section.

The experimenter has, after all, used only wires of metal *A*. He now proceeds to the level of confirmation and speaks to himself thus: If my (great) discovery is to be made public, I have to check it, using other metals as well. I shall do the last step of the experiment, namely, find resistance values for different diameters of wires of *B*, *C*, and *D*, plotting in each case resistance against the area of cross-section, to see if I can confirm a similar relation. This is hypothesis (v). So, now he performs his planned actions with anxiety. His theory passes the test: with confirmation, his hypothesis is now raised to the level of a theory.

The following points are significant to us in this narration:

1. A problem is the starting point of an investigation.
2. The problem is of interest—in fact, meaningful—only to those who are familiar with the background from which the problem arises.
3. The hypothesis is a tentative answer to the problem at hand (expressed in the form of an assertion).
4. The hypothesis must be confirmed or rejected based on experiment (or available information—the theory in the field).
5. The hypothesis precedes the experiment.
6. In some cases, experimental findings lead to better or different hypotheses.
7. There may be several iterations of the hypothesis–experiment phase.
8. The hypothesis, confirmed by experimental findings, becomes the theory.
9. A theory needs further confirmation; in this sense, the new theory takes the place of the hypothesis.

10. Based on this theory, now confirmed, further experiments or observations can be decided; these are often called *crucial experiments* (or *tests*).
11. If crucial experiments confirm the theory, it is established as a piece of scientific knowledge.

No theory, however well established, is the last word. If and when it fails to pass some test, either in a new experiment or implied by a new theory, it is open for reconsideration, modification, and even rejection.

6.2 Desirable Qualities of Hypotheses

The use of hypotheses are as widespread and their varieties are large. Isaac Newton's Laws of Motion were hypotheses he set himself to prove. Finding my lost set of keys also requires hypotheses. Some qualities serve as criteria to distinguish great hypotheses from trivial ones and, to an extent, to judge the soundness of a given hypothesis. Hypothesizing is part of human thought—we all do it in our daily lives. But to produce a hypothesis of the stature of the Laws of Motion is an extraordinary event, even among great scientists.

All those who have a Ph.D. in philosophy are not philosophers; all those who have a Ph.D. in literature are not poets or writers. (In fact, great creative writers seldom have Ph.D's) But most scientific researchers—experimental or theoretical—have had the experience, however slightly, of forming scientific hypotheses. In fact, in the course of research, let us say the three years spent by a Ph.D. candidate, the most crucial time, the turning point, is the event of formulating the hypothesis. Once that is done, in a sense, the rest is routine: doing experiments, writing the theses, receiving the degree, and, of course, looking for a job.

The excitement of incubating and hatching a hypothesis is quite a feeling; it has come, and it is gone! During that period or soon after, it is worthwhile for the researcher to spend some time analyzing the hypothesis itself to make sure that it is not a fake. In these days of group research—with the emphasis on teamwork—it is possible that the individual researcher will be given a hypothesis and asked to use his skill and knowledge to do the experiments, then create a report or a thesis. Such an individual is lucky in that he avoids a lot of frustration, akin to searching for an unrecognizable object in a dark room, and dead-end literature studies. But he is also deprived of the worthwhile excitement of forming the hypotheses in his head and carrying it through to confirmation, working like an artist, giving shape to his dream!

Whether a group or an individual forms the hypothesis, subjecting the hypothesis to scrutiny is desirable. The following are some well-known qualities of a good hypothesis:

1. The first and foremost purpose of a hypothesis is to answer the specific question at hand as closely and directly as possible, without bringing into the domain of inquiry unnecessary details and impertinent connections. For example, if our investigator on electrical resistance set himself the task of shining the wires to absolute cleanliness or finding the metallurgical composition of each wire to the accuracy of the third decimal, he might be perfecting some possible variables, but he would be leading himself away from the problem at hand.
2. A hypothesis should be testable, and the results of the test should be observable in terms of material objects and entities, as physical or chemical attributes or measurable qualities or changes, which can be directly perceived. If a psychic person claims that he can see ghosts and tells us their size, but asserts that others cannot see them and that the size cannot be found by ordinary measuring instruments, then the existence of ghosts, stated as a hypothesis, is useless because nobody can confirm this assertion by means of experiments.
3. A hypothesis should have deductive capacity. Suppose a hypothesis is confirmed by means of a set of tests or experiments; the confirmed hypothesis then becomes a part of the body of science. This fact should not exhaust the capacity of the confirmed hypothesis (from then on referred to as a theory) to deduce one or more consequences of

itself that will also be true. Each such consequence, in effect, becomes a new hypothesis to be tested. To the extent that a deduced hypothesis also passes the test, the original hypothesis gains credibility. A well-known example in recent times is the confirmation test done for the Theory of Relativity. A deduction from the theory was that light, like any material object, is subjected to gravitation. Though the theory was presented in 1905, it was not fully accepted by the scientific community until the test was done in 1919, and the result was found to be positive. The greater the deductive capacity of a given hypothesis, the better the hypothesis. We should note, however, that deductive capacity may appear similar to, but is quite different from, testability.

4. A new hypothesis should be consistent with the existing, established body of science in that field. Most often the contribution made by new research, say doctoral work, is like adding another brick to the growing wall for a room in a big building under construction. The brick should "form a good fit" to be acceptable and useful as a modest part of the building. But there are exceptions to this rule; nonconformity is not necessarily always a defect. Very rarely, it may happen that a hypothesis, by not fitting well, will call for an entirely new structuring of science—a whole new building! For example, the principle of the conservation of mass, which is still taken for granted in most physics, chemistry, and technology calculations, had to be given up on strict consideration when mass and energy were shown to be equivalent in the early part of this century. Now, we have, instead, the principle of the conservation of energy, wherein mass is simply a component. Is the new conservation principle true for all time? So far, yes, but there may be nothing that lasts for all time in science.

5. A simple hypothesis is better than a complex one. As an artist could say, "Simplicity is beauty"; so could a scientist. But the answer to the question, What is simplicity? may not be obvious. The artist may venture to explain: instead of painting a young female model wearing elaborate clothing and jewelry, he prefers to paint her nude because the latter arrangement, to him, is simple. But it may not be so easy for a scientist to find an equivalent "simple" arrangement to convince himself that his hypothesis is simpler than a rival hypothesis. Simple testability (meaning the requirement of a simple and direct test or experiment for confirmation), the capacity to beget one or more deductions that can in turn be used as crucial tests, and the ability to explain one or more prevailing puzzles in science are some attributes of a simple, and superior, hypothesis.

6.3 Bibliography

Choen, M. R., and E. Nagel. *An Introduction to Logic and Scientific Method.* Boston: Rutledge and Kegan Paul, 1961.

Greenwood, D. *The Nature of Science.* Wisdom Library, 1959.

Poincare, H. *Science and Hypothesis.* Mineola, NY: Dover Publications, 1952.

7

Designing Experiments to Suit Problems

A scientific or technical study always consists of the following three steps:

One decides on the objectives.

One considers the method.

One evaluates the method in relation to the objective.

—*Genichi Taguchi*

An experiment is a means to an end. The end is an answer to a question that the experimenter formulates relative to a chosen, finite part of nature. There are several kinds of experiments, each suited to the question to be answered. Within the framework of cause-effect relation, "one-cause, one-effect" questions are by far the easiest to formulate. But starting with some of the experiments devised to answer such questions, this chapter looks at the logical and practical inconsistencies that arise. Experiments suitable to answer a few circumstantial variations in cause-effect relations are briefly described. Further, we point out that one-cause, one-effect experiments, besides being wasteful, are most often logically unjustifiable. The discussions in this chapter lead us to experiments involving the combined effects of more than one cause, a subject dealt with in Chapters 8 and 9.

7.1 Several Problems, Several Causes

The association of one cause with one effect has historically been considered "obvious"; hence, the logic of *Mill's Methods of Inquiry* (see Chapter 4). Boyle's law (1662), for example, associates the pressure (P) of a gas with its volume (V): $P \propto 1/V$. Another law, attributed to Charles (1746–1823) and Gay (1778–1850) Lussac, associates pressure (P) with (absolute) temperature (T): $P \propto T$. It is implied, though, that in the verification of $P \propto 1/V$, the temperature is to be maintained constant, and in the verification of $P \propto T$, the volume is to remain unaltered. If, for instance, we consider the experiment in which the effect of pressure on volume is being verified, with increasing pressure, the temperature concomitantly increases; to keep the temperature constant, the heat generated needs to be removed concurrently. To make this happen, the systems of insulation, temperature monitoring, and heat removal have to be perfect and instantaneous, which is, in terms of experimental control, unattainable. The laws were formulated disregarding such realities, that is, by "idealizing" the experiments. In terms of systems associated with an effect driven by several causes simultaneously, the above example is only the tip of the iceberg.

Instances where close to a dozen or more causes simultaneously determine the end effect are not rare. The yield of an agricultural crop and the quality of a manufactured piece of hardware are but two examples. That yield in agriculture is a result of many variables—seed quality, field preparation, rain fall, sunshine, fertilizer, pest control, and so forth—functioning together, has long been known to the farmer. To bring the benefit of scientific research to agriculture, experiments entailing many causes acting simultaneously are required. R. A. Fisher and his contemporary investigators, starting in the early twentieth century, initiated a systematic study with the application of statistical principles to such experiments, which, in fact, laid the foundation for later analysis of experiments involving more than one factor. Many more vigorous developments have since been taking place in extending this aspect of statistics to experimental research in such areas as medicine, psychology, education, manufacturing, athletics, and so on. Such developments came to be collectively known as "Design of Experiments."

Though Design of Experiments has found applications in diverse areas, it is very often expounded in terms of plants and agriculture, perhaps owing to its origin. But analogies have been found, and extensively used, in all other fields of application. Conforming to this practice, Design of Experiments can be broadly divided into two aspects: (1) designing the *block structure*, and (2) designing the *treatment structure*.

The first of these, the *block structure*, refers to dividing a certain area of an experimental site for agriculture into so many *blocks*, and each block into so many *plots*, the purpose being to assign each plot to a different *treatment*. The second, the *treatment structure*, consists of subjecting different varieties of plants to the effect(s) of a combination of more than one cause, each cause referred to as a *factor*. Both structures are created using well-known principles of statistics and in ways that conform to *randomization*. Block structure and randomization are discussed in Chapter 16. The design of treatment structures, or *factorial design*, is discussed in this chapter. A few terms, often used as equivalents owing to their entry from different fields of application, need to be mentioned:

- *Factor*: also referred to as a cause, input, treatment, parameter
- *Result*: also referred to as an effect, yield, response, output, quality characteristic, dependent variable
- *Treatment*: also referred to as a given combination of factors at given levels
- *Test*: also referred to as trial, replication, experiment, and so forth

The design referred to, obviously, is the preparatory work of the experimenter, in purpose and function similar to a blueprint for an engineer before execution of a project. The basic question that the experimenter needs to ask in designing an experiment is, What question(s) should the result(s) of the test(s) help me to answer? A careful, detailed answer to such a question determines what treatment—inclusion and/or variation of one or many independent variable(s), to act as cause—is required This constitutes the treatment structure, which has several variations; some of the better-known ones are briefly described as follows.

7.2 Treatment Structures

7.2.1 Placebo

Placebo means something to please. It is not rare that some patients experience relief from a sickness on using a medicine given by a doctor, even if the medicine is nothing more than an inert substance given to the patient, without his knowledge, in the form (shape and color) of a medicinal unit. I remember a doctor who practiced in a rural area. Some of her patients walked miles to come to her with complaints of various sicknesses: stomachache, giddiness, weakness, headache, and so forth. They insisted on having an injection instead of tablets, even though they knew the doctor would charge them less if they accepted tablets. The doctor had tubes of distilled water waiting to be used for such patients, who paid her gladly after receiving the injection. In either case, tablets or injection, patients were often given a placebo.

Imagine that a drug company comes up with a proprietary product to cure one such sickness, and it falls to the lot of an experimenter to justify or repudiate the claim of the company. If two patients, otherwise comparable, are experimented on, one is given the proprietary product, and the other is given a fake, a placebo, that looks similar. If both of the patients are cured equally well, the proprietary product's worth is called into question. If, instead, the one who used the proprietary product is cured and the other one is not, the product is worthy to that extent.

7.2.2 Standard Treatment

Suppose two plants of a kind, which seem equal in every way, are subjected to test. Let us say that one of these is given a plant food. To the extent that this plant gives a better yield, the plant food is considered beneficial.

7.2.3 "Subject-and-Control" Group Treatment

Instead of only two plants as mentioned in the previous section, two groups of plants, otherwise comparable, are subjected to test for the benefit of plant food. One group is subject to treatment, and the other is not. Yields of the two groups, analyzed on a statistical basis, are then compared; to the extent that the group that received the treatment gives a better yield, the plant food is rated as beneficial.

7.2.4 Paired Comparison Treatment

A particular plant food is, let us say, claimed to benefit apple, pear, and orange trees in terms of yield. One member of each pair of comparable apple, pear, and orange trees is given the plant food, while the other member of each pair is denied it; both members of each pair are otherwise provided similar care. The yield for each kind of plant is recorded separately. This data provides the basis to justify or repudiate the claim that the plant food can benefit these kinds of plants and further to quantify the benefit by statistical methods.

7.2.5 Varying the Amount of One of the Two Factors

In the simple case of one cause, one effect, if the cause is varied over a reasonable range, and the corresponding effects are recorded, we have the classic x-y relation that can be plotted as a graph. If, instead, two factors are known to be influential, and one is kept constant while the other is varied, we get a "surface" that can be plotted as a graph. Here, if x_1 and x_2 are the factors (causes), with x_1 remaining constant and x_2 varying, and y is the result (effect), we obtain a surface of width x_1 and shaped according the x_2-y relation.

7.3 Many Factors at Many Levels, but One Factor at a Time

Let us first illustrate a case in which two or more factors have a combined influence but are experimented with one factor at a time.

Example 7.1

We call our experimenter here "coffee lover." He guessed that "great" coffee is the result of adding to the fresh brew of a given strength of a given brand of coffee the right quantity of cream and the right quantity of sugar, then drinking it at the right temperature. But, to his disappointment, the experience of "great" coffee did not happen every day. On those rare days that it happened, he secretly said to himself, "Today is my lucky day." Inclined to be scientific, one day he decides to find the right combination and be done with the luck business. After considerable reflection on his coffee taste, he decides to vary the cream between 0.5 and 2.5 teaspoons in steps of 0.5 teaspoons; to vary the sugar between 0.5 and 2.5 teaspoons in steps of 0.5 teaspoons; and to start drinking after 0.5 and 2.5 minutes in steps of 0.5 minutes, after preparing the cup of coffee, steaming hot.

He symbolized the order of increasing quantities of the variables as below:

Cream	c_1	c_2	c_3	c_4	c_5
Sugar	s_1	s_2	s_3	s_4	s_5
Time	t_1	t_2	t_3	t_4	t_5

Further he writes down the following combinations:

$c_1s_1t_1$	$c_2s_1t_1$	$c_3s_1t_1$	$c_4s_1t_1$	$c_5s_1t_1$
$c_2s_2t_2$	$c_3s_2t_2$	$c_4s_2t_2$	$c_5s_2t_2$	$c_1s_2t_2$
$c_3s_3t_3$	$c_4s_3t_3$	$c_5s_3t_3$	$c_1s_3t_3$	$c_2s_3t_3$
$c_4s_4t_4$	$c_5s_4t_4$	$c_1s_4t_4$	$c_2s_4t_4$	$c_3s_4t_4$
$c_5s_5t_5$	$c_5s_1t_1$	$c_5s_2t_2$	$c_5s_3t_3$	$c_5s_4t_4$

He sets to work methodically to drink one cup of coffee at the same hour on consecutive days for each of the above combinations. During this period, of over three weeks, he often has to gulp some pretty bad stuff because those were not the right combinations for his taste. Only on a few days does he find the taste to be somewhat like, but not exactly, the right taste he knows. He is beginning to feel the fatigue of the experiment. At the end of those twenty-five days, still unsatisfied, he takes a second look at the experimental schedule he prepared and followed through.

He notices the following:

1. There are a lot more possible combinations than his table included.
2. The first line of his table gives the effect of varying c, but does so with s, and t, both at the lowest levels only. Even if he finds the best level for c, it is possible, indeed more than likely, that the levels s_1 and t_1 are not the best.
3. The second line also gives the effect of variation for c, but again is tied with only one level each of s and t, creating the same doubt as before.
4. The first column of the table, and partly the other columns also, show the effect of varying c, but mixed up with different s-t combinations. He had wanted to keep the s-t combination constant. The levels are s_2t_2, s_5t_5, and so forth, always incrementing together. Where is, for example, a combination like s_2t_4 or s_3t_5?
5. Where is a column or a line that shows the effect of varying s only, keeping the c-t combination constant?

Our coffee lover is now fatigued, confused, and frustrated. He fears that with all these goings on, he may develop an aversion for coffee; that would be a terrible thing to have happen. Not committed to secrecy in experimenting, he explains the situation to a friend, who happens to be a math teacher. After a few minutes of juggling figures, the teacher tells him that there are one hundred more possible combinations that he could add to his table of experiments. Noticing disappointment written large on his friend's face, the teacher tells him that this experiment format, to be called *one factor at a time*, is not the best way to go and offers to help him devise a shorter and more fruitful program if and when his friend is ready. The coffee lover takes a rain check, having decided that coffee is for enjoying, not for experimenting—at least for now. Having witnessed the frustration of the one-factor-at-a-time experiment, we may now illustrate a simple case of factorial design.

7.4 Factorial Design, the Right Way

When two or more factors, each in two or more levels, are to be tested for their combined effect on the quality characteristic—the dependent variable—*factorial design* is appropriate. The following is an example of three factors at two levels.

Example 7.2

In a sheet metal press, it is the intention to measure the quantity of deformation of a metal sheet with various combinations of three factors: press pressure (p), duration of press action (d), and temperature of plate (t). If each factor is to be tried at two levels, (p_1, p_2), (d_1, d_2), and (t_1, t_2), we have the following eight treatments with different settings of factors:

(1) $p_1d_1t_1$ (2) $p_2d_1t_1$ (3) $p_1d_2t_1$ (4) $p_2d_2t_1$

(5) $p_1d_1t_2$ (6) $p_2d_1t_2$ (7) $p_1d_2t_2$ (8) $p_2d_2t_2$

Comparing (1) and (2) above, we notice that d_1t_1 is common between the two, and there is change in p alone; p functions as an independent variable (cause) with consequent change in the dependent variable (effect) observed or measured and recorded. Combination pairs (3)-(4), (5)-(6), and (7)-(8) also show p in the role of independent variable, with other factors remaining same. Together, we have four cases in which the effect of changing the parameter, p (the cause), from level p_1 to p_2 can be observed. Each of these is analogous to the paired comparison tests we discussed earlier; instead of the factors being present or absent completely, we now have the factor functioning at two levels, p_1 and p_2.

Similar cases can be made for the parameters d and t also; the summary is shown in Table 7.1.

Instead of the above scheme, which is the essence of factorial design, if we resorted to a one-factor-at-a-time schedule to obtain, as above, four comparisons of each of the three factors, each at two levels, we would require $4 \times 3 \times 2 = 24$ treatments, and that too, not as appropriately as the eight treatments discussed above.

Table 7.1 *Effect of changing the parameters from low level to high level.*

Cause (factor functioning as a parameter, changing from "low level" to "high level")	Effect (factor combinations numbered below, serving as four paired comparisons to study the effect of the factor)			
p	1–2	3–4	5–6	7–8
d	1–3	2–4	5–7	6–8
t	1–5	2–6	3–7	4–8

7.5 Too Many Factors on Hand?

Efficient as it is, even factorial design can get choked with too many treatments in industrial experiments, where it is not uncommon to face as many as ten to fifteen factors threatening to act simultaneously on the outcome.

Consider an experiment in which there are six factors—this is not too many in several industrial contexts—and where it is the intention to find the optimum level of each. It is decided to study each factor at four levels. The number of replications to run all possible combinations is $N = (4)^6 = 1{,}024$. Obviously, this is too many replications. Also, when a "nuisance factor" (described later in this chapter) is suspected to be influential as a cause, the remedy is to design the experiment to treat such a factor as yet another factor, in par with the usual factors. This, in effect, adds to the number of treatments needed, $N = l^{(f)}$, exponentially!

Several means, based on statistical methods, have been developed to reduce the number of replications to a lower, reasonably convenient level. A design using all combinations given by $N = l^{(f)}$ is known as a *full factorial*, whereas, designs using only a part of such combinations are often done so to reduce the number of treatments—are referred to as *fractional factorials*. The method of using fractional factorials in industrial manufacturing has been adapted extensively by Genichi Taguchi and others.

7.6 "Subjects-and-Controls" Experiments

Situations wherein several causes acting together result in one or more noticeable effects are not rare. Our coffee lover's experiment was but one example. If the intention of the experiment is to study the effect of a new cause in addition to the existing ones, then a comparison between two cases, one with the additional cause and another without it, all other conditions being similar, becomes necessary. The study may pertain to objects, animate or inanimate, or to nonobjective phenomena; each item in the study is referred to as an *individual.* The normal procedure in such a study is to

1. Select a fairly large, reasonable number of individuals by a random method, the individuals representing the population.
2. Divide these by a random method into two groups of equal number.
3. Apply to the first group all the usual causes.
4. Apply to the second group all the above causes, plus the additional *cause,* to find the effect of which forms the experiment.
5. Observe the result, the effect(s), in the first group as a statistical average.
6. Observe the effects in the second group as a similar, statistical average.
7. Compare the two results obtained above.

The difference in effect(s) in the groups is attributed to the additional cause. In such experiments, individuals in the first group are known as *controls*, and those in the second group are known as *subjects.*

Example 7.3

Suppose it is the intention of an experimenter to try the effect of a new plant food on the yield of a certain variety of annual flowering plant. Then, the experimenter randomly selects, let us say, one hundred seedlings from a larger lot. He plants these in one hundred pots, one plant in each, identical in all respects: size, material, soil, moss, manure, fertilizer, moisture, and so forth. He marks by a random method fifty of these as subjects. The other fifty plants become controls. He cares for all one hundred pots with identical gardening procedures, such as space, sunshine, warmth, watering, pesticide, and so on. He does not distinguish one in any way from the others in his routine care all through their growth until the flowering time, with the only difference that he adds the dose or doses of experimental plant food uniformly to all subjects in predetermined quantities and on a predetermined schedule. In time, the plants begin flowering. He records the yield from each subject and arrives at a statistical average for all subjects. Whether the "yield"—the criterion—is greater number of flowers, bigger flowers, brighter flowers, or flowering early in the season or any other single or combination of criteria is either predetermined or left open by the experimenter to determine later. Similarly, he obtains a statistical average for the controls. The effect of the experimental plant food may be beneficial, deleterious, or neutral. This is decided by comparing the two statistical averages. If the subjects yielded more flowers, for example, the effect of the new plant food is beneficial to that extent.

7.6.1 Varieties within Subjects and Controls: Paired Comparison Design

Example 7.4

Let us imagine that an investigator in the gardening experiment was very pleased with the efficacy of the plant food as he has a stake in its promotion. He wants to extend his investigation to know if many other kinds of flowering plants benefit as well from the plant food. Let us call the particular variety of annuals he has recorded his success with as *A* and the other varieties as *B*, *C*, *D*, *E*, *F*, *G*, *H*, and *J*. Now he plans his experiment differently. He selects two plants of *A* at random from a larger lot. After considering the features and properties that he knows are significant in deciding plant fitness or healthiness, leading to yield, and after analyzing such features and properties, he is convinced that the two plants of *A* are of equal merit. Following the same procedure, he selects two plants each of *B*, *C*, *D*, *E*, *F*, *G*, *H*, and

J. Between the two plants of *A*, he decides, by tossing a coin, which one of the two will be the subject and which the control. Let us now call the subject A^1 and the control *A*. Along the same lines, he decides between B^1 and *B*, C^1 and *C*, and so forth, and, accordingly, tags the pots containing the plants. This procedure of selecting two individuals that are as close as one can find in regard to the discernible features and properties that make the two individuals of equal merit or potentiality is known as *pairing*. As described above, the purpose is to use one as the subject and the other as the control. Our experimenter has now on hand eighteen pots of plants: *A, B, C, D, E, F, G, H,* and *J* to be used as controls and A^1, B^1, C^1, D^1, E^1, F^1, G^1, H^1, and J^1 to be used as subjects.

At this stage, we will assume that the routine treatments—pot size, the kind and quantity of soil and moss, the amount and frequency of watering, weather control, including need for pesticides, and so forth—required for *A* and A^1 are different from those required for *B* and B^1 and *C* and C^1, and so on. All these treatments—and we assume our experimenter knows these procedures well—are met as required. All the routine treatments given to *A, B, C, . . . J,* are also given correspondingly to A^1, B^1, C^1, . . . J^1. The one and only difference is that plant food, not given to *A ,B, C, . . . J,* is given to A^1, B^1, C^1, . . . J^1 as an "extra treatment." The "dosage" appropriate for *A, B, C, . . . J* is decided on a somewhat arbitrary basis but subject to the expertise of the experimenter. Such routine cares as are considered appropriate are given through the flowering periods. Using the criterion of "improvement," the yield of A^1 is compared with that of *A*, the yield of B^1 with that of *B*, the yield of C^1 with that of *C*, and so forth. At this time, we assume that each subject's yield was better than that of the corresponding control, and to that extent our experimenter has reason to be convinced that the experimental plant food benefits "all" annual flowering plants. If the question arises in his mind about the degree of benefit, then he needs to quantify his criteria and follow the appropriate numbers in terms of statistical analysis, which we will deal with in Chapter 19.

7.6.2 Experiments with Humans

Investigations like that above, known as *paired comparison design*, are fairly common in medical research, when experimenting with drugs and procedures. The problem of designing the experiment in such studies becomes more complex for the following reasons:

1. When dealing with humans, it is difficult to find identical patients. Age, race, sex, heredity, physical features, educational level, social background, philosophical attitudes, habits, and temperament are some of the factors that distinguish individuals, and these factors have significant influence on reactions and responses to treatments. It is no exaggeration to say that there are no two exactly identical persons. Identical twins may be closest to the ideal, but to find sufficient numbers of such twins, simultaneously available as "patients," is extremely rare. Otherwise, forming several pairs, one half of which is to serve as control and the other half as subject, is fairly common; the choice of who the subject will be is determined by the toss of a coin. With several such pairs, a set of controls and another set of subjects, can be obtained; all this means that the population from which the individuals can be selected for the experiment is very large and the search correspondingly difficult.
2. Humans, unlike plants, have rights. It is often necessary to obtain informed consent from the individuals in an experiment. If some individuals do not consent, the required population needs to become even larger.
3. Even after the consent is obtained, the decision, though made purely by chance, that one of the pair will receive the treatment and the other will not has ethical implications. If the treatment is known to be beneficial, the subject individuals stand to gain. If the treatment turns out to be harmful, the subjects are victimized, with the questionable justification that there is a greater

good beyond the individuals involved. If Immanuel Kant could breathe, he would shout from his grave that this justification is improper. Is not each individual an end to himself, not just a means? Industrial products, at the other extreme, fortunately for the experimenter involved, do not pose such disturbing questions.

7.7 Combined Effect of Many Causes

In the preceding example of an experiment on the benefit of a new plant food, it was agreed by implication that the addition of the plant food to the nourishment protocol of a subject plant did not, by its presence, in any way influence the other items of care. Suppose that a particular plant's physiology was such that to digest the experimental food, it needed an extra supply of water. This was denied to the plant on the grounds that the quantity and schedule of watering should differ from those used for the control plants. If the need for the extra supply of water were known beforehand and were provided, the benefit of the plant food could be even greater. On the other hand, because this need was denied, there could even be a deleterious effect instead of a gain. In paired comparison experiments of medicines on humans, such "unknown" effects can be even more pronounced. If some of the subjects, for instance, develop an allergy to the new drug, while other subjects benefit remarkably, the inference from the experiment will be inconclusive. These hypothetical, but possible, cases have this in common: several causes operate simultaneously and yield a combined and summary effect. When such effect is to be observed or quantified, the relative significance of each cause needs to be analyzed in the context of its being one of many causes, not acting independently, and with the possibility of influencing and being influenced by other causes in its contribution toward the summary effect. Many successful attempts to face such problems in agriculture, biology, medicine, education, sports, engineering, and other sciences are collectively known as *factorial design of experiments*; this topic is dealt with in Chapters 8 and 9.

7.8 Unavoidable ("Nuisance") Factors

In the context of many factors acting together, some unintentional, often unperceived or unavoidable, factors that cannot be accounted for may influence, to an unknown extent, the quality characteristic. For example, vibrations in the building, noise, illumination, room temperature, humidity, and so forth, can have an effect when the experiment involves only inanimate objects. When animals are involved, more "biological" factors appear, and when humans are involved, more "human" factors are added to the list. Collectively, these are known as *nuisance* or *noise factors*. There are three aspects to the existence of such factors:

1. Their influence on the quality characteristic or summary effect, acting alone or with other factors, may be ignorable or considerable in extent.
2. The experimenter's awareness of these factors may vary from vague suspicion to real concern.
3. When there is concern, the experimenter's ability to compensate for them may vary from total helplessness to complete control.

Confined to the "triangle" of these three sides, figuratively speaking, nuisance factors have sway in most experimental research. Fortunately, some control is possible, whether noticed or not, by the experimenter, by means of *randomization*, which is dealt with in Chapter 16.

7.9 Bibliography

Anderson, D., D. J. Sweeny, and T. A. Williams. *Introduction to Statistics: An Applications Approach.* St. Paul, MN: West Publishing Co., 1981.

Fisher, R. A. *Statistical Methods, Experimental Design and Scientific Inference.* Cambridge, UK: Oxford University Press, 1990.

Groninger, L. D. *Beginning Statistics with a Research Context.* New York: Harper and Row Publishers, 1990.

Wilson, E. B., Jr. *An Introduction to Scientific Research.* Mineola, NY: Dover Publications, 1990.

8

Dealing with Factors

We have usually no knowledge that any one factor will exert its effects independently of all others that can be varied, or that its effects are particularly simply related to variations in these other factors.

—Sir Ronald A. Fisher

This chapter introduces factorial experiments. When two or more factors are known to act together as independent variables, one-factor-at-a-time experiments are shown to be inadequate. The concept of "experimental space" for three-factor, two-level experiments is demonstrated. When more than three factors are involved, the concept requires more than three dimensions. This chapter demonstrates how this is a limitation and how matrices can be used instead. Beginning with two factors, we elaborate the method for planning any number of factors, all at two levels, and further for analyzing the main effects of, and interactions between, the factors. The use of a fractional factorial with five factors, all at two levels, is shown, with an example of handling a large number of factors. The final part of the chapter summarizes several variations of factors and the forms in which they may prevail in factorial experiments.

8.1 Designing Factors

The most striking feature that distinguishes experiments with designed factors from those without them is that, in the former, all the factors are simultaneously designed into the experiment, whereas in the latter, experiments are done one factor at a time, like experiments with only one independent variable. That designing factors is more efficient than conducting one-factor-at-a-time experiments is the first point we need to underscore. The number of many factors and number of levels at which each should be tested, of course, depend on the particular experimental requirement. For our purpose, we will start with three factors, each to be tested at two levels. We will call the factors a, b, and c. When used at low levels, we will designate these as a_1, b_1 and c_1, and at high levels, as a_2, b_2 and c_2. Functions of the three independent variables, when all are used at low levels, may be expressed as

$$y_1 = f(a_1, b_1, c_1)$$

and when all are used at high levels, as

$$y_2 = f(a_2, b_2, c_2)$$

In this designation, "high" and "low" do not necessarily mean that $a_2 > a_1$, with both expressed in the same units. Instead, it means that by changing a from a_1 to a_2, an *improvement* is expected in the dependent variable y. Improvement, again, does not mean that $y_2 > y_1$, with both expressed in the same units. For instance, if y is the profit in a business, when y_2 is numerically higher than y_1, there is an improvement, whereas if y is the loss in the business, when y_2 is numerically higher than y_1, there is not an improvement; it would then be desirable that y_2 be numerically less than y_1. Similar considerations hold good for the independent variables b and c as well.

8.2 Experiments with Designed Factors

Consider the following eight experiments with combinations:

(1) $a_1b_1c_1$ (2) $a_2b_1c_1$ (3) $a_2b_2c_1$ (4) $a_1b_2c_1$

(5) $a_2b_1c_2$ (6) $a_1b_1c_2$ (7) $a_1b_2c_2$ (8) $a_2b_2c_2$

Out of these, each of the following four pairs provides results for comparing the effect of a_2 with that of a_1:

(1) $a_1b_1c_1$	with	(2) $a_2b_1c_1$
(4) $a_1b_2c_1$	with	(3) $a_2b_2c_1$
(6) $a_1b_1c_2$	with	(5) $a_2b_1c_2$
(7) $a_1b_2c_2$	with	(8) $a_2b_2c_2$

The other four pairs serve to compare the effect of b_2 with that of b_1:

(1) $a_1b_1c_1$	with	(4) $a_1b_2c_1$
(2) $a_2b_1c_1$	with	(3) $a_2b_2c_1$
(5) $a_2b_1c_2$	with	(8) $a_2b_2c_2$
(6) $a_1b_1c_2$	with	(7) $a_1b_2c_2$

Finally, the following four pairs provide the bases to compare the effect of c_2 with that of c_1:

(1) $a_1b_1c_1$	with	(6) $a_1b_1c_2$
(4) $a_1b_2c_1$	with	(7) $a_1b_2c_2$
(2) $a_2b_1c_1$	with	(5) $a_2b_1c_2$
(3) $a_2b_2c_1$	with	(8) $a_2b_2c_2$

Thus, having done only eight *replications* with different combinations, it is possible to compare the effects of a, b, and c, each at two different levels, with each comparison evidenced four times. Yet another point is highly worthy of note. The effect of changing, for instance, a from level a_1 to a_2 is available for our observation in the presence of four different combination of factors, namely, b_1c_1, b_2c_1, b_1c_2, and b_2c_2. And, when we summarize the main effect of a by averaging the four changes in the dependent variable, we are, in effect, getting to study the behavior of a

in the company of all the possible associates. Needless to say, the same advantage applies relative to factors *b* and *c*.

On the other hand, a one-factor-at-a-time experiment for studying the effect of *a*, as an independent variable, consists of conducting two replications—one at each level of *a*, namely, a_1 and a_2, keeping the levels of *b* and *c* constant in both these experiments. And similarly, when *b* and *c* are independent variables, the other factors are kept constant. Let us say that the following combinations of *a*, *b*, and *c* were experimented on:

1. a_1, b_1, c_1 and a_2, b_1, c_1 for the effect of *a*

   Suppose the combinations with a_2 were found to yield better effects than those with a_1; from then on, a_1 would be dropped and no longer included in any further combinations.

2. a_2, b_1, c_1 and a_2, b_2, c_1

   If the combination with b_2 were found to yield a better effect, b_1 would not be included in any further combinations.

3. a_2, b_2, c_1 and a_2, b_2, c_2

   If the combination with c_2 were found better, c_1 would be dropped.

In the above experiment scheme, in which a_2 is better than a_1, b_2 is better than b_1, and c_2 is better than c_1, each of these decisions is based on the strength of just one piece of evidence made possible by one pair of replications. Besides, this arrangement studies the "behavior" of the factors, each *in only one company of associates.* But as all experimental researchers know, or would be made to know, the outcome of a single experimental comparison as the means of evidence cannot be depended on. To gain the same level of confidence as in the scheme of experiment with designed factors, even ignoring the other possible association of factors, we need to replicate four times each pair of trials. That means the total number of trials is $2 \times 3 \times 4 = 24$. Thus, the information provided by eight trials with designed factors requires twenty-four trials by the scheme of a one-factor-at-a-time experiment.

8.3 Matrix of Factors

The situation of having three variables, each at two levels, can be represented pictorially, as shown in Figure 8.1. Each factor combination of the eight combinations we have dealt with as "designed factors" is represented by a vertex of the orthorhombic volume; the space represented by this shape is often referred to as the *experimental space.* If each factor has more than two levels, the experimental space can be extended, and the various factor combinations can be visually represented. If more than three factors are involved, however, then we need a space of more than three dimensions, which cannot be represented visually. That is the limitation of the experimental space. Statisticians have devised several other ways of compiling the factor combinations, which do not suffer such limitations; these are often referred to as *matrices.* The experiments involved are referred to as *matrix-design* experiments, also more popularly known as "factorial-design."

Figure 8.1 *Experimental space.*

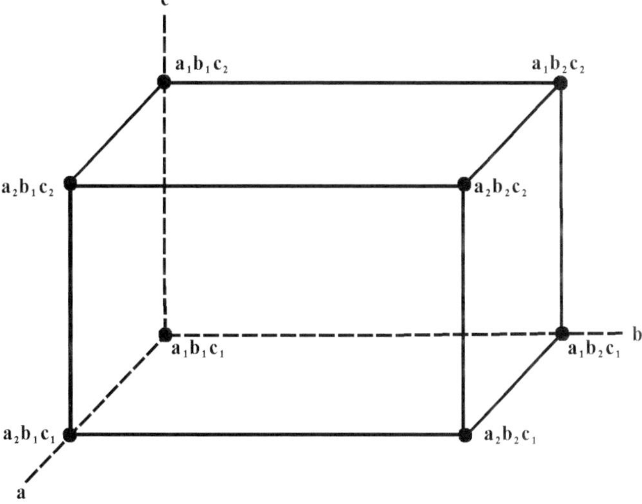

1. When only two levels of factors are involved, the words "high" and "low" can be used to represent any number of factors. Thus, the eight combina-

Table 8.1 *Three Factors at Two Levels Each*

Replication	Factors		
	a	b	c
1	low	low	low
2	high	low	low
3	high	high	low
4	low	high	low
5	high	low	high
6	low	low	high
7	low	high	high
8	high	high	high

tions we dealt with earlier can be represented as they are in Table 8.1.

2. The arithmetic symbols "+" and "–" are often used in place of "high" and "low," respectively, making the matrix look more crisp, as shown in Table 8.2.

Table 8.2 *Three Factors, Two Levels with "–" and "+" Symbolism*

Replication	Factors		
	a	b	c
1	-	-	-
2	+	-	-
3	+	+	-
4	-	+	-
5	+	-	+
6	-	-	+
7	-	+	+
8	+	+	+

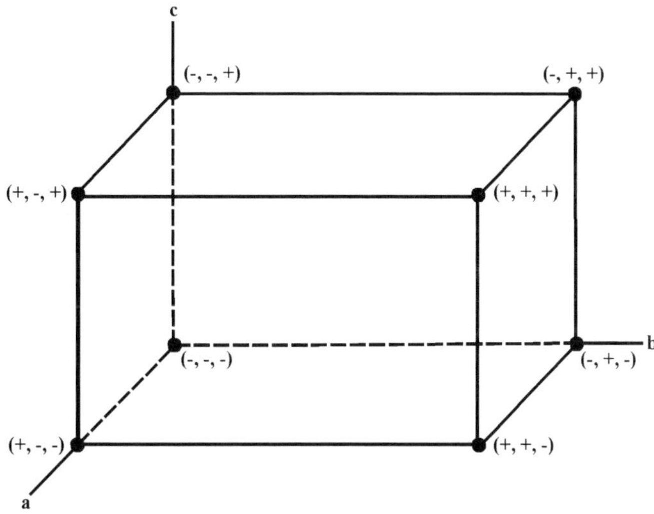

Figure 8.2
Experimental space with "+" and "–."

If we adapt this symbolism, the experimental space shown in Figure 8.1 can be represented as shown in Figure 8.2.

3. A slight modification of the above method is to use "+1" in place of "+" and "–1" in place of "–"; this makes it possible to determine the *interaction of factors* given by such arithmetic relations as (+ 1) × (+ 1) = + 1, (–1) × (–1) = +1, (+1) × (–1) = –1, and (–1) × (+ 1) = –1. We will see the convenience of this symbolism further in this chapter.

4. Another way of symbolizing the experimental space is to identify each combination with only those letters that are "high," and to leave out those that are "low" in the combination. According to this symbolism, referring to Figure 8.1,

   $a_1b_1c_2 \equiv c,\ a_1b_2c_2 \equiv bc$

   $a_2b_1c_2 \equiv ac,\ a_2b_2c_2 \equiv abc$

   $a_2b_1c_1 \equiv a,\ a_2b_2c_1 \equiv ab$

   $a_1b_2c_1 \equiv b$

The combination $a_1b_1c_1$, instead of being blank, is represented by "1." In graphical form, it can be represented as shown

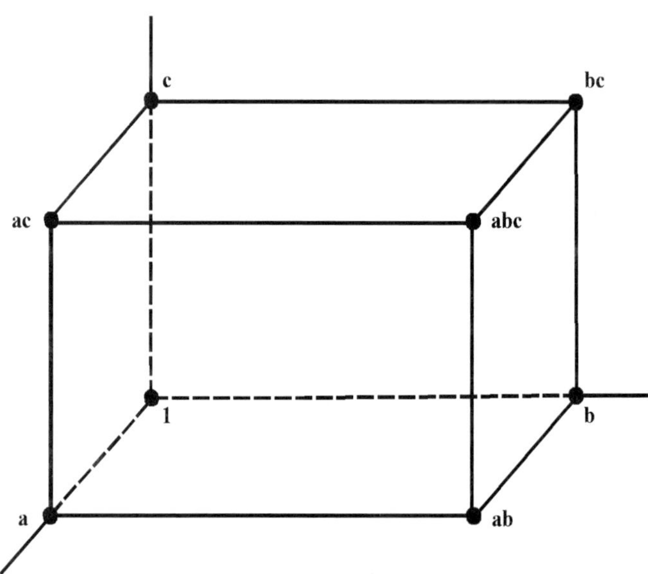

Figure 8.3 *Experimental space with single-letter symbolism.*

in Figure 8.3. In tabular form in Table 8.3, we can notice that this matrix offers an interesting relation, which will be even more obvious when dealing with a larger number of factors.

Table 8.3 *Three Factors, Two Levels with Single-Letter Symbolism*

Combination	Can be symbolized as	No. of Combinations
$a_1b_1c_1$	1	1
a_2,b_1,c_1	a	as many as there are variables
$a_1b_2c_1$	b	3
$a_1b_1c_2$	c	
$a_2b_2c_1$	ab	n = 3; r = 2
$a_1b_2c_2$	bc	${}^nC_r = 3$
$a_2b_1c_2$	ca	

Table 8.3 *Three Factors, Two Levels with Single-Letter Symbolism (continued)*

Combination	Can be symbolized as	No. of Combinations
$a_2b_2c_2$	abc	1
	Total:	8

8.3.1 More Than Three Factors

Suppose there are four variables: *a*, *b*, *c*, and *d*. Using this new symbolism, we may list the combinations as shown in Table 8.4.

Table 8.4 *Three Factors, Two Levels with Bifurcation of Factor Levels*

Number	Combination Symbol	Feature of combination	Actual combination	No. of combination given by	No. of combinations
1	1	All *low*	a_1, b_1, c_1, d_1	Only one such	1
2 3 4 5	*a* *b* *c* d	One symbol for each variable	a_2, b_1, c_1, d_1 a_1, b_2, c_1, d_1 a_1, b_1, c_2, d_1 a_1, b_1, c_1, d_2	Number of variables	As many as there are variables: 4
6 7 8 9 10 11	ab bc cd da bd ac	Combination of two symbols	a_2, b_2, c_1, d_1 a_1, b_2, c_2, d_1 a_1, b_1, c_2, d_2 a_2, b_1, c_1, d_2 a_1, b_2, c_1, d_2 a_2, b_1, c_2, d_1	n = 4 r =2 ^nC_r	6
12 13 14 15	abc bcd cda dab	Combination of three symbols	a_2, b_2, c_2, d_1 a_1, b_2, c_2, d_2 a_2, b_1, c_3, d_2 a_2, b_2, c_1, d_2	n=4 r=3 ^nC_r	4
16	1	All high	a_2, b_2, c_2, d_2	Only one such	1
				Total:	16

However, the reader should note that the case of four factors at two levels each cannot be represented graphically as an experimental space because it requires four dimensions; nonetheless, it can be expressed as a matrix. Further, the matrix method, on the same lines as shown in Table 8.4, can be used for five, six, indeed, any number of variables. It is also capable of providing the number of combinations under each category of one, two, three, or more symbols representing the combinations of variables. It is obvious that all these forms of matrices are suitable for situations wherein factors are tried at only two levels. Matrices suitable for three or more levels for each factor will be dealt with in Chapter 9.

We will conclude this section with some more terminology used in the context of designing factors. Both the eight-trial experiment with three variables and the sixteen-trial experiment with four variables, wherein each factor was listed for its effect at two levels, are known as *complete* (or *full*) *factorial experiments*, implying that all possible combinations of factors are tested. It is also to be noted that the three-factor, two-level experiment has eight possible combinations. These numbers are related as $2^3 = 8$. This experiment, hence, is referred to as a *2^3 experiment*. Along similar lines, the four-factor, two-level situation is referred to as a *2^4 experiment* and has $2^4 = 16$ possible combinations. Further, this relation can be generalized for any complete (or full) factional experiment: the number of combinations of factors, possible = l^f, wherein

l = number of levels at which each factor is tried

f = number of factors involved

A full-factorial experiment with four factors, each at five levels, according to the above relation, requires $5^4 = 625$ replications. In such situations, a more reasonable, smaller number of replications can usually be justified and is accomplished by eliminating some factors, some levels, or some of both. Such multilevel-multifactor experiments, with replications less than *full factorial*, are referred to as *partial* (or *fractional*) *factorial experiments*.

8.4 Remarks on Experiments with Two-Level Factors

In multifactor experiments, particularly when the number of factors is high, testing at two levels of factors is quite adequate to decide (1) if one or more of the factors is ineffective, (2) the relative extent of effectiveness of each factor, and (3) if two or more of the factors *interact*, meaning that their combined effect is not merely cumulative. We have seen that the number of trials for three factors at two levels in full-factorial experiments is given by $2^3 = 8$. If the number of factors is increased to four, the total number of trials is $2^4 = 16$, which is a big jump from eight. There are several situations in which close to a dozen or so factors can be identified as likely to be influential. In such cases, a full-factorial experiment, even with only two levels per factor, can become unmanageable.

A well-known method for dealing with problems in industrial manufacturing is popularly known as the Taguchi method. It consists of listing for a specified response all the possible factors; this is done by a whole team of experts drawn from all fields—not just engineering—involved in the product. Then, in a brainstorming session—this is important—in which people of diverse ranks and job specialties participate, each of the factors is scrutinized, some are rejected by joint discretion, and the list of factors is considerably shortened. Using only these final factors, a full-factorial, two-level experiment is carried out. After analyzing the response and identifying and rejecting those factors determined to have marginal influence, the list of factors is further shortened. With the remaining short list of factors, a partial-factorial, multi-level-multifactor experiment is scheduled.

8.5 Response of Multifactor Experiments

Whereas results of single-factor or one-factor-at-a time experiments can usually be compiled as $y = f(x)$, the response to multi-factor experiments needs a different treatment for analysis. We will start in this chapter with the simplest situation: two factors, at two levels each. We further assume that the factors are quantitative. But to keep the analysis adaptable to several different situations, we will

1. Not identify the factors by their properties or dimensional units
2. Identify the factors only by names, as *a*, *b*, *c*, and so forth
3. Identify the levels of the factors by one of the many ways that we have mentioned earlier, namely, either as "low" and "high"; "–" and "+"; or a_1 or a_2 for *a* (and similarly for *b*, *c*, and so on), as this is convenient when there are only two levels
4. Record the responses as hypothetical quantities (for illustration) devoid of dimensional units

Table 8.5 shows the replication results of all combinations possible (full factorial) with two factors, *a* and *b*, with *a* at two levels, a_1 and a_2, and *b* at two levels, b_1 and b_2. The number of combinations is given by $\ell^f = 2^2 = 4$.

Table 8.5 *Two-Factor, Two-Level Experiment with (Hypothetical) Responses*

Replication	Factor/Levels *ab*		Response
1	- (*low*, a_1)	- (*low*, b_1)	36.4
2	+ (*high*, a_2)	- (*low*, b_1)	35.8
3	- (*low*, a_1)	+ (*high*, b_2)	42.3
4	+ (*high*, a_2)	+ (*high*, b_2)	31.9

Figure 8.4 is the graphical presentation of the results shown in the second column of Table 8.5 for the change in *a*. The reader should also note that while the low level of *a* (a_1) is shown first, then the high level (a_2), from left to right on the abscissa, there is no need to have a scale, and thus, one is not used. On the other hand, a scale is required for the ordinates, as these measure the *responses* in quantities as given in the last column of Table 8.5. Restricting our focus to Figure 8.4, we can measure the *main effect* of *a* as follows: factor *a* is changed from low (a_1) level to high (a_2) level, once with b_1 and another time with b_2, each of these changes having caused a difference in the response. Consid-

ering that there are two such changes, the effect of a (changing from a_1 to a_2) is given by the mean of the two differences: (response with a_2 minus response with a_1). That is,

$$\text{Effect of } a = [(35.8 - 36.4) + (31.9 - 42.3)] \div 2 = -5.5$$

The negative sign, of course, should not be a surprise. As we mentioned earlier, the response could be a "loss" in a business; the negative sign then is an "improvement"! This same numerical value, intact with the sign, can be obtained if we now associate the "–" and the "+" signs in the a column of the table (the second column) with the corresponding numerical values of the response column (the fourth column). That is,

$$\text{Effect of } a = [-36.4 + 35.8 - 43.3 + 31.9] \div 2 = -5.5$$

Herein lies the advantage of using "–" and "+" signs instead of "low" and "high" or a_1 and a_2, respectively. On the same lines as Figure 8.4, we can now represent the effect of b on the responses. The same four numbers in the last column of Table 8.5 now get assigned to "low" and "high" in the b (the third) column of the table. The result is shown in Figure 8.5. The effect of b can now be calculated, as shown above, by simply assigning the "–" and "+" signs of the b column in Table 8.5 to the corresponding numerical values for response in the last column. Thus,

$$\text{Effect of } b = [-36.4 - 35.8 + 42.3 + 31.9] \div 2 = 1.0$$

Before we proceed further to deal with more involved situations, we need to take note of two definitions, related to the graphical presentations in Figures 8.4 and 8.5.

1. The effects we have computed above for factors a and b are known as the *main effect* because each refers to the primary function of that factor as an independent variable, somewhat as if it were the only factor acting upon the dependent variable.

Figure 8.4 *Experimental outcome with two factors at two levels; effect of b.*

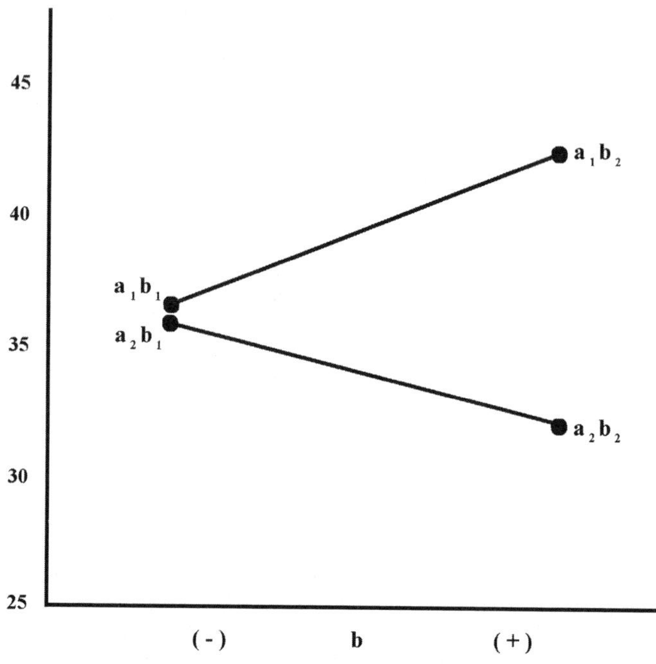

Usually, the symbols a and b are also used for the *main effects* of the corresponding factors as well.

2. In either figure, the line showing the effect of a and that showing the effect of b are nonparallel. In Figure 8.4, the lines actually cross each other. Such a situation is referred to as *observation of an interaction* between the two factors a and b. In Figure 8.5, though the lines are nonparallel, there is no *interaction* within the experimental limits of the factors. If the two lines in Figure 8.4 were parallel, there would be no chance of their crossing each other, whatever the experimental limits of the variables. This situation would be referred to as *no interaction.*

Now, what is the significance of *interaction* between the two factors a and b? It is an indication that a given factor fails to produce the same effect on the response at different levels of the other factor. In Figure 8.4, for instance, at low levels of b (b_1), the effect of a is

$$a = 35.8 - 36.4 = -0.6$$

and at high levels of b (b_2), the effect of a is

$$a = 31.9 - 42.3 = -10.4$$

This means that the effectiveness of a as an independent variable is sensitive to the amount of b that is coexisting and acting as another independent variable.

The extent of interaction, where it exists, is measurable. In our designation in Figure 8.4, it is referred to as the *a • b interaction* and is often symbolized as *ab*. The magnitude of the interaction, by definition, is half the difference between the two effects of *a:* the first at a high level of b (b_2) and the second at a low level of b (b_1). That is,

$$ab = [(31.9 - 42.3) - (35.8 - 36.4)] \div 2 = -4.9$$

As a cross-check, suppose we now calculate the *ba* interaction, meaning the change in the effectiveness of b with the presence of a at different levels. At a high level of a (a_2),

$$b = (31.9 - 35.8) = -3.9$$

and, at a low level of a (a_1)

$$b = 42.3 - 36.4 = 5.9$$

Then,

$$ba = [-3.9 - 5.9] \div 2 = -4.9$$

As is to be expected, the extent of interaction is the same; that is, $ab = ba$.

We have shown earlier that multifactor experiments are more efficient than one-factor-at-a-time experiments. Here is yet another advantage of multifactor experiments: the aspect of interaction between factors could never be traced in a one-factor-at-a-time experiment.

8.6 Experiments with More Factors, Each at Two Levels

As a way of preparing for factorial experiments with more than two factors, let us slightly modify the data in Table 8.5, using the "–1" and "+1" symbolism in place of "low" and "high" for the factor levels. The modification results in Table 8.6.

Table 8.6 *Two-Factor, Two-Level Experiments with "–1" and "+1" for Factor Levels*

Replication	Factor/Levels a	b	Response
1	-1	-1	36.4
2	+1	-1	35.8
3	-1	+1	42.3
4	+1	+1	31.9
	A = -11.0	B = 2.0	A • B = -9.8
Main Effects and Interaction	A/2 = -5.5 a	B/2 = +1.0 b	(A • B)/2 = -4.9 ab

An explanation of the last two rows of this table is needed.

Evaluation of *A* was done as follows: Multiply each "response" value as found in the last column by either "–1" or "+1" as found in the *a* column in the corresponding row of the table. Then, add the resulting values together.

$$A = [36.4(-1)] + [35.8\ (+1)] + [42.3\ (-1)] + [31.9(+1)] = -11.0.$$

Similarly,

$$B = [36.4(-1)] + [35.8\ (-1)] + [42.3(+1)] + [31.9(+1)] = +2.0.$$

The value of $A \bullet B$ is found on similar lines, with this difference; each response value is multiplied by the product of either "+1" or "–1" as found in column *a* and either "+1" or "–1" as found in column *b*. Sum of the four products thus obtained, shown below, is the value of $A \bullet B$.

$$A \bullet B = [36.4(-1)\ (-1)] + [35.8\ (+1)\ (-1)] + [42.3(-1)\ (+1)] + [31.9(+)\ (+1)] = -9.8$$

The values of *A*, *B*, and $A \bullet B$ were derived using all four response values. The experiment yielding these responses could be formulated as $y = f(a, b)$, wherein *y* gives the responses (the dependent variables), and *a* and *b* are two factors (independent variables) acting together. The effect of *a* on *y*, for instance, was tested for at two levels, a_1 and a_2, each requiring one replication. This means that two replications were necessary, yielding responses, to study the effect of *a* once. Four responses tested, thus, gives the effect of *a*, $4 \div 2 = 2$ times. So, the main effect of *a* is obtained by dividing *A* by 2. The last row of Table 8.6, thus, shows the main effect of *a*, that of *b*, and the interaction of *a* and *b*, respectively, symbolized as *a*, *b*, and *ab*.

The advantage of using the "–1" and "+1" symbolism for levels of factors must now be obvious to the reader. Evaluation of the main effects and the interaction can be done "mechanically," without the need to penetrate the logic, and avoiding the possible confusion involved.

Now we may consider a *full factorial* of three factors, *a*, *b*, and *c*, each at two levels. The number of combinations is $2^3 = 8$, as shown in Table 8.7.

At this stage, the experimenter is expected to conduct the experiment with eight different factor combinations and to get a response for each combination. Table 8.8 shows the above combinations symbolized by "–1" and "+1," respectively, for the

Table 8.7 *Three Factors, Two Levels, Eight Combinations*

1.	low a	low b	low c
2.	high a	low b	low c
3.	low a	high b	low c
4.	high a	high b	low c
5.	low a	low b	high c
6.	high a	low b	high c
7.	low a	high b	high c
8.	high a	high b	low c

Table 8.8 *Three-Factor, Two-Level Replications with (Hypothetical) Responses*

Replication	a	b	c	Response
1	-1	-1	-1	38.2
2	+1	-1	-1	40.1
3	-1	+1	-1	40.8
4	+1	+1	-1	46.7
5	-1	-1	+1	39.2
6	+1	-1	+1	43.4
7	-1	+1	+1	37.6
8	+1	+1	+1	50.0

"low" and "high" values of the three factors, with the corresponding hypothetical numerical values of the responses.

Among *a*, *b*, and *c*, we get the interactions *ab*, *bc*, *ac*, and *abc*. We may now extend the above table to include a column for each of these four interactions. As for the "–1" and "+1" signs required corresponding to each of these columns, a simple multiplication rule is all that is needed. For instance, in the third row of Table 8.8, we find *a* with "–1," *b* with "+1," and *c* with "–1" signs. On the same row, thus, *ab* will get a "–1" sign, as $(-1) \times (+1) = -1$; *bc* will also get a "–1" sign, as $(+1) \times (-1) = -1$; *ac* will get a "+1"

Table 8.9 *Three-Factor, Two-Level Experiments with Interactions and (Hypothetical) Responses*

	a	b	c	ab	bc	ac	abc	Response
1	-1	-1	-1	+1	+1	+1	-1	38.2
2	+1	-1	-1	-1	+1	-1	+1	40.1
3	-1	+1	-1	-1	-1	+1	+1	40.8
4	+1	+1	-1	+1	-1	-1	-1	46.7
5	-1	-1	+1	+1	-1	-1	+1	39.2
6	+1	-1	+1	-1	-1	+1	-1	43.4
7	-1	+1	+1	-1	+1	-1	-1	37.6
8	+1	+1	+1	+1	+1	+1	+1	50.0

sign, as (–1) × (–1) = +1; and *abc* will get a "+1" sign, as (–1) × (+1) × (–1) = +1. Working out such details can be done without

Table 8.10 *Main Effects and Interactions Calculated*

Factor Combination	a	b	c	ab	bc	ac	abc	Response
1	-1	-1	-1	+1	+1	+1	-1	38.2
2	+1	-1	-1	-1	+1	-1	+1	40.1
3	-1	+1	-1	-1	-1	+1	+1	40.8
4	+1	+1	-1	+1	-1	-1	-1	46.7
5	-1	-1	+1	+1	-1	-1	+1	39.2
6	+1	-1	+1	-1	+1	-1	-1	43.4
7	-1	+1	+1	-1	+1	-1	-1	37.6
8	+1	+1	+1	+1	+1	+1	+1	50.0
Sum	24.4	14.2	4.4	12.2	-4.2	8.8	4.2	
Divisor	4	4	4	4	4	4	4	
Contrast:	← Main effects →				← Interactions →			
	6.1 3.55 1.1				3.5 -1.5 2. 2 1.05			

the need for any further information. The data in Table 8.9 shows this method applied to all the eight rows.

Using this table as the source, let us follow the method for calculating the *main effects* or *interactions*, together known as *contrasts*, each represented by a column within the matrix. We will do it for *a* to find the *main effect* of that factor. Following the procedure used with two factors in Table 8.6, now extending to three factors, we have

A = [(–1) (38.2) + (+1) (40.1) + (–1) (40.8) + (+1) (46.7) + (–1) (39.2)+ (+1) (43.4) + (–1) (37.6) + (+1) (50.0)] = 24.4. The sum, 24.4, represents four responses at "low" levels and four responses at "high" levels of *a*. As two responses together, one at "low" and one at "high" levels of *a*, are required to know the effect of *a* once, the sum, 24.4, needs to be divided by four to get the "average effect" of *a*. The values of such sums for all seven columns, and the values of contrasts, obtained by dividing each by four (referred to as the *divisor*), are shown in Table 8.10.

Thus far, we have shown how three factors may be represented. We will now go one step further and demonstrate how a matrix can be generated to represent four factors, *a*, *b*, *c*, and *d*, each at two levels. Then, the number of factor combinations is given by $2^4 = 16$. Table 8.11 shows the combinations and interactions, together constituting the matrix, in terms of "–1" and "+1."

The "pattern" procedure required to generate the matrix for even larger numbers of factors must now be obvious to the reader. We point out a few noticeable features:

1. The column corresponding to the first factor has "–1" and "+1" alternating one by one as we go down, factor combinations one to sixteen. (Note: There is no need to prioritize or establish an order among factors; they may be taken in any order.)
2. Such alternating for the second factor/column is two by two; for the third factor/column it is four by four; for the fourth factor/column it is eight by eight.

Table 8.11 *Four-Factor, Two-Level Experiment with Contrasts*

Factors Combination No.	a	b	c	d	ab	bc	cd	ad	bd	ac	abc	bcd	abd	acd	abcd	Experimental Response
1	-1	-1	-1	-1	+1	+1	+1	+1	+1	+1	-1	-1	-1	-1	+1	?
2	+1	-1	-1	-1	-1	+1	+1	-1	+1	-1	+1	-1	+1	-1	-1	?
3	-1	+1	-1	-1	-1	-1	+1	+1	-1	+1	+1	+1	+1	-1	-1	?
4	+1	+1	-1	-1	+1	-1	+1	-1	-1	-1	-1	+1	-1	+1	+1	?
5	-1	-1	+1	-1	+1	-1	-1	+1	+1	-1	+1	+1	-1	-1	-1	?
6	+1	-1	+1	-1	-1	-1	-1	-1	+1	+1	-1	+1	+1	+1	+1	?
7	-1	+1	+1	-1	-1	+1	-1	+1	-1	-1	-1	-1	+1	+1	+1	?
8	+1	+1	+1	-1	+1	+1	-1	-1	-1	+1	+1	-1	-1	-1	-1	?
9	-1	-1	-1	+1	+1	+1	-1	-1	-1	+1	-1	+1	+1	-1	-1	?
10	+1	-1	-1	+1	-1	+1	-1	+1	-1	-1	+1	+1	-1	+1	+1	?
11	-1	+1	-1	+1	-1	-1	-1	-1	+1	+1	+1	-1	-1	+1	+1	?
12	+1	+1	-1	+1	+1	-1	-1	+1	+1	-1	-1	-1	+1	-1	-1	?
13	-1	-1	+1	+1	+1	-1	+1	-1	-1	-1	+1	-1	+1	-1	+1	?
14	+1	-1	+1	+1	-1	-1	+1	+1	-1	+1	-1	-1	-1	-1	-1	?
15	-1	+1	+1	+1	-1	+1	+1	-1	+1	-1	-1	+1	-1	-1	-1	?
16	+1	+1	+1	+1	+1	+1	+1	+1	+1	+1	+1	+1	+1	+1	+1	?
sum	?	?	?	?	?	?	?	?	?	?	?	?	?	?	?	?
Divisor	8	8	8	8	8	8	8	8	8	8	8	8	8	8	8	?
Contrasts	← Main Effect →				← Interaction →											
	?	?	?	?	?	?	?	?	?	?	?	?	?	?		

Based on these observations, the pattern can be formulated as shown in Table 8.12.

For the four columns in Table 8.11, each one representing one factor, the *interactions* are obtained by multiplying the corresponding "–1" and "+1" symbols, representing the factor levels, along the rows (across the columns). The number of interactions

Table 8.12 *Assigning "–1" or "+1" in a Multifactor Experiment*

Column/ factor number	Number of '-1's' then '+1's' (in bunch) alternating
1	$2^0 = 1$
2	$2^1 = 2$
3	$2^2 = 4$
4	$2^3 = 8$
5	$2^4 = 16$
6	$2^5 = 32$
7	$2^6 = 64$
...	...
n	2^{n-1}

between any two factors, given by the number of combinations of two factors at a time among the four factors is

$$^{n}C_{r} = \frac{n!}{r!\times(n-r)!}$$

$$= \frac{4!}{2!\times 2!} = 6$$

and these are *ab*, *bc*, *cd*, *ad*, *bd*, and *ac*. The interaction among these factors, given by the combination of three factors at a time among four factors is

$$^{4}C_{3} = 4!/3! = 4$$

These are *abc*, *bcd*, *abd*, and *acd*. The interaction among all four factors is, of course, given by one combination, *abcd*. Thus, there are 4 + 6 + 4 + 1 = 15 columns, each yielding a contrast, four contrasts of main effects, and eleven contrasts of interactions. With the four combinations shown in Table 8.6, there are three contrasts. With the eight combinations in Table 8.10, there are seven contrasts. With the sixteen combinations in Table 8.11, there are fifteen contrasts. We may, by induction, generalize that the number of contrasts will be equal to the number of combinations, minus one.

The reader should notice here that in each column, there are as many "–1" symbols as there are "+1" symbols. This feature follows in any matrix, however big it may be.

Now, if we have the numerical values of the responses obtained by experimentation, corresponding to each of the sixteen factor combinations, we can create a column next to the matrix, as shown in Table 8.11. The values of the main effect of the four factors and eleven interactions, together with fifteen contrasts, can each be evaluated simply by adding the sixteen products obtained by multiplying "–1" or "+1" in a given row of the column to the corresponding response value on the same row. In place of each response and each contrast, the values of these being unknown, a "?" mark is registered.

8.7 Fractional Factorials

Having arrived at the logical end point for discussing full-factorial experiments with four factors, we need to reflect on what lies ahead. We mentioned early in this chapter that in many contexts, for example, manufacturing, confronting a large number of factors is not uncommon. If the number of factors is ten, for instance, even at only two levels (let alone three and more levels), with full factorial, we have 2^{10} = 1,024 factor combinations! Obviously, there ought to be some shortcut to avoid the need to run such a large number of replications. A selected small number of combinations, referred to as a *fractional factorial*, are often found adequate for the initial screening of the factors. We mentioned this earlier, while referring to Taguchi.

Example 8.1

This is a problem faced by a metal-casting company. One of the requirements of molding sand is to permit the gases to escape to reach the atmosphere. This requirement is known in the industry as *permeability*. Many factors influence permeability. Some of the factors, their levels, and their effects in quantities are discussed with an example in Chapter 9. In this chapter, to demonstrate the use of fractional factorials, with each factor at only two levels, we assume that there are only five factors; they are here designated by *a*, *b*, *c*, *d*, and *e*. It is known that a *balanced factorial*, also often referred to as an *orthogonal array*, can be formed with the number of test runs, one more than the number of factors. Table 8.13 shows such a balanced factorial for the five factors with only six corresponding responses, y_1, y_2, y_3, y_4, y_5, and y_6, one for each factor combination. In passing, we note that a full factorial of five factors, instead, would have required $2^5 = 32$ test runs.

Table 8.13 *Fractional Factorial of Five Factors*

Test Run #	Factor a	Factor b	Factor c	Factor d	Factor e	Response/ Dependent Variable
	Factor Levels					
1	1	1	1	1	1	y_1
2	1	2	2	1	2	y_2
3	1	1	2	2	2	y_3
4	2	2	1	1	2	y_4
5	2	1	2	2	1	y_5
6	2	2	1	2	1	y_6

The "balance" referred to above, besides other aspects, can thus be noticed in the matrix.

1. In each column under the factors, there are as many level 1's as there are level 2's; this is also true for the entire matrix.

2. Except for the first line, which is used as the reference for the "now" condition, all other lines have the same number of level 1's (2) and level 2's (3).

Using these values of dependent variables, shown only by the symbols y_1 to y_6 here, we can track the main effect of each factor, while acting in combination with other factors. The difference between the average of the values of dependent variables for a given factor when that factor is at level 2, and such an average when that factor is at level 1, is the measure of the main effect of that factor. For instance, the main effect of *c* is given by

$$\frac{y_2 + y_3 + y_5}{3} - \frac{y_1 + y_4 + y_6}{3}$$

or $$\frac{y_2 + y_3 + y_5 - y_1 - y_4 - y_6}{3}$$

In the context of this experiment, if the above value is positive, it indicates *c* as a beneficial factor. Similar values calculated for the other four factors will give us, on a comparative basis, the degree of effectiveness of all the factors. From these, we may identify the most effective one (or two) of the factors. For further optimization, one may plan for either a two-factor, two-level experiment or a two-factor, three-or-more-level experiment, as required. Tracing the interactions (not done here) between those two factors will then be fairly easy.

8.8 Varieties of Factors

8.8.1 Quantitative versus Qualitative Factors

For convenience, we have so far considered a factor as an independent variable in a function of the kind $y = f(x)$. In this equation, for every different numerical value of x, there is a corresponding numerical value of y. Therefore, x, then, is obviously a quantitative factor. The effect of temperature on a chemical reaction, the effect of the dosage of a drug on the response of patients, and the effect of carbon percentage on the hardness of steel are some examples of such relations. Though this is the most obvious form of a factor, factors are manifested in several other forms in experiments. The sexes of humans in a psychological test, for instance, are two levels of one qualitative factor, human. Strains of laboratory animals, varieties of wheat in a test for yield, and light bulbs made by different companies subjected to test for length of life are other examples of qualitative factors.

8.8.2 Random versus Fixed Factors

The worth of the published results of an experiment is measured by the extent to which generalization is possible. If the findings are restricted to the particular variety of: subject or material, time of day, and place of experimentation, then the factors so involved are said to be *fixed.* To facilitate generalization, there should be randomization relative to such factors. If, for instance, an experiment is planned to test the effect of the dosage of a pain-killing drug on human adults, ideally, randomization should be imposed relative to the sex, age, race, social, and professional varieties of the subjects, the body locations of pain, and time of day, and several replications should be done over a lengthy period in different weather and climate conditions, and so on. Out of these and many others conceivable factors, only those on which randomization was imposed in the experiment are referred to as *random* factors; those that were not so randomized remain *fixed* factors in the experiment.

8.8.3 Constant and Phantom Factors

As mentioned above, potential factors are many; in fact, the experimenter him- or herself is a potential factor; the room where the experiment is done is also a potential factor. In any experiment, only a limited number of potential factors can be used as design factors. The remaining ones can be classified into two types: *constant* factors and *phantom* factors. Constant factors are those used at only one level throughout the experiment. When all experiments are done in one room, the room is a constant factor; if all replications are done by one experimenter, the experimenter is a constant factor. Phantom factors, on the other hand, are those that enter the experimental system at different levels, though unintentionally and often uncontrollably. Time of the day is a good example. Change in room temperature and humidity (when these are not included as design factors) are more examples. The obvious remedy against the effects of both the constant and phantom factors is randomization.

8.8.4 Treatment and Trial Factor

Treatment in our context means "to find the effect(s) of." If, for instance, performance of a car in terms of miles per gallon is being tested for at speeds of fifty, sixty, and seventy miles per hour, then the speed, which is purposely varied, is a *treatment factor*. In the case of multifactor experiments, in which the effect of each factor, over a range, on the outcome is tested for, each is referred to as a treatment factor. When the experimenter is not sure whether a certain factor has significant or negligibly small effect on the outcome, that factor will be designed into the experiment in the preliminary stage as a treatment factor. If it is found to be an insignificant factor, from then on, it will be treated in further experiments either as a fixed or constant factor, depending on the context. In contrast, when some factors—with the levels of each factor held steady—are repeatedly tested for their combined effects, and the outcome of each repetition is recorded to be averaged later for better confidence, each factor is called a *trial factor* (and such tests are referred to as *repeated-measurement experiments*).

8.8.5 Blocking and Group Factors

Blocking refers to dividing subjects (before the experiment) into different groups on the basis of similarity within each group and distinction from other groups. *Blocking factors* may be either qualitative or quantitative. Some instances follow: If humans (or any other animals) are to be experimented on, blocking them into "male" and "female" groups is qualitative. Occupation and ethnic origin are other examples. If adult men are the subjects, blocking by age, say twenty to forty, forty-one to sixty, sixty-one to eighty years, is quantitative. Years of college education and yearly income are other examples. In contrast, forming groups among subjects by arbitrary assignment leads to *group factors*. Suppose a given drug is to be tested on humans for its effect at a given dosage and also at twice that dosage with a subject population of twenty adults. Forming two groups of ten with randomization and testing the single dosage on one group and the double dosage on the other does not constitute a group factor. On the other hand, suppose a class of ten students is arbitrarily divided into two groups of five (say, those in the first bench and those in the second bench), and each group is required as a team to work out a solution on a collaborative basis to the same problem given to both the groups. A talented individual in one group or a disruptive individual in the other is enough to make a considerable difference between the two groups' performances. As a way of being fair to the students, this possibility calls for a group factor: students included in the group deprived of the proven advantage need to be given an extra boost in evaluation.

8.8.6 Unit Factor

The individuals in a subject group are referred to as *units*. If units in two or more groups are all subjected to the same test, and the score of each unit is separately recorded, it is quite possible, even likely, that the differences in scores will be considerable. Then, the average of scores within a particular group serves as an index of that experiment. Such averages are the bases for *unit factors*.

These are some, but by no means all, of the variations of factors. By now, it should be evident to the reader that what makes a given factor *random* (not *fixed*), *constant* (not *phantom*), or *treatment* (not *trial*), and so on, is not the nature of the factor as such but the circumstances that bring the factor into the experimental scheme, as perceived and decided upon by the experimenter. Based on the answers to such questions as the following, a set of factors will first be selected by the experimenter:

- What are all the potential variables that may influence the effect to be studied?
- Which of these variables are to be included—now termed factors in the experiment—and which, if any, may be ignored?
-

Each of these will, in turn, be subjected to more intrusive questions, such as

- Is this factor of direct interest as a cause? If the answer is yes, it will become one of the treatment factors.

- May this factor modify the effect of one or more of the treatment factors? If the answer is yes, the presence or absence of interaction between factors will be focused on.
- Is this factor a part of the experimental technique (e.g., varieties of rice in the study of the effect of a given fertilizer on the yield of rice)?
- Is this factor required because of the need for classification (e.g., men and women in an experiment on the effect of a given drug for headache relief)?
- Does the presence or absence of this factor make a difference in the effect to be studied (e.g., a certain vitamin proposed to have catalytic value when taken with a given miracle weight-loss drug)?

As the experimenter answers these questions, each selected factor will fall into any one, but usually more than one, of the various classification(s) discussed above.

8.9 Levels of Factors

Considerations, of quantity or quality levels, is an inseparable part of the selection of factors. First of all, the difference relative to a given variable may be taken either as a distinction between (or among) factors or as different levels of one factor. Men and women, in an experiment dealing with physical strength, may be two levels on one factor, humans. In another experiment dealing with family expenses for clothing and cosmetics, it is prudent to think of the two, men and women, as different factors; this decision, of course, is left to the experimenter.

The presence or absence of a particular variable may be considered as two levels of one factor. When the factor is quantitative, absence may be treated as level zero. When the factor is qualitative, there is room for confusion. For instance, it is often claimed that plants are sensitive to classical music. I am not aware of any proof for or against such a claim, but I know of a high school English teacher who, while his students were taking an examination, played records of classical music, believing in a similar claim that human performance is affected by music. If that teacher had two groups of; say; thirty students separated with randomization, and if he placed each group in identical rooms and gave all individuals the same examination at the same time, with himself functioning as proctor for both groups with equal time sharing, and if, with these factors common, he played music in one room and let silence prevail in the other, music then would be a presence-or-absence qualitative factor. Using the same criteria for judging, if the average performance of those in the group that was exposed to music was found to be better, to that extent, the belief of the teacher, the "hypothesis" of that "experimenter,"—would be proven valid. But then, it would be a typical one-factor-at-a-time experiment, which is less meaningful. To observe the effect of music as a factor properly, the extents of its interaction(s) or noninteraction(s) with other factors, such as difference(s) in the subject matter for examination, teacher personality, time of day, duration of the examination, and so on, need to be studied.

Assuming that a list of factors has been selected, we now look at the selection of levels separately for quantitative and qualitative factors.

8.9.1 Levels of Quantitative Factors

Quantitative factors need to be specified in terms of

1. The lowest and the highest measurable value
2. The number of steps between these extremes
3. The increments (or spacing) between consecutive steps

In these respects, each factor in factorial experiments requires similar considerations to the ones in one-factor-at-a-time experiments.

The selection of extreme values is sometimes obvious. For instance, in an experiment in which temperature is a factor in measuring the productivity of shop workers, it is reasonable not to go beyond the range of 40°F to 100°F. On the other hand, if the experiment is to determine the effectiveness of a weight-increasing wonder drug, the lowest level should, of course, be zero, but the selection of the highest level is not so obvious. The guiding principles in selecting the number of steps are, on the one hand, to take as many as possible within the resources, but, on the other hand, to take as few as are sufficient for economy. In preliminary experiments, it is reasonable to use a considerably smaller number of levels than in the final experiment. Also, if the effect is known to present a linear relation, a smaller number of levels, often only two, is adequate, whereas if the effect presents a curve instead, at least three levels for the preliminary experiment, and many more for the final experiment, are required. As for the space between levels, it is preferable in most cases to keep it uniform, but if the output presents a combination of straight and curved parts, closer spacing at the curve than at the straight part is desirable.

8.9.2 Levels of Qualitative Factors

An obvious example of a qualitative factor is different varieties of rice in testing the effectiveness of a fertilizer. The levels are the same as the varieties tested; hence, their number is the same. Another example is the case discussed above using classical music as a factor. Varieties in the music delivered based on volume, composer, instrument, and so on, can become qualitative factors with as many levels as there are variations. In both of these examples, the highest and lowest levels, as well as the spacing between them, are devoid of significance.

8.10 Bibliography

Antony, Jiju. *Design of Experiments for Engineers and Scientists.* Boston: Butterworth Heinemann, 2003.

Cox, D. R. *Planning of Experiments.* New York: John Wiley and Sons, 1992.

Grove, Daniel, and Timothy P. Davis. *Engineering, Quality and Experimental Design.* Essex, UK: Longman Scientific and Technical, 1992.

9

Factors at More Than Two Levels

The aim of science is not things themselves, as the dogmatists in their simplicity imagine, but the relation between things.

—*Henri Poincaré*

Factorials at three or more levels are necessary to trace the relation between the independent variables acting together and the response, the dependent variable. Starting with a two-factor, four-level experiment, in this chapter identifying the interactions is graphically demonstrated. Measuring the main effects and interactions by means of two-way tables is shown. A practical problem is taken from the metal-casting industry as an example. Selecting four out of many factors, using the Quality Circle procedure, and adapting a three-level fractional factorial, the steps taken toward the solution are discussed in detail.

9.1 Limitations of Experiments with Factors at Two Levels

Chapter 8 dealt with designing many factors together, all at only two levels. When we are beset with a large number of factors, it is possible that one or more among those are spurious. It is desirable to eliminate those early in planning the experiment. If the main effect of a particular factor is zero or close to zero, it should draw our attention as a possibly ineffective factor. We need to see if that factor has any noticeable interaction with other factors, particularly with the prime factor (assuming that it is possible, either by previous experience or by intention, to designate one as such). If the interaction is also negligible, the particular factor should be counted out in all further steps of the design. The more such spurious factors we can identify, the less complex the planning and analysis of the responses from the experiment becomes. Two-level experiments are ideally suited for that. Besides, data from two-level experiments provide an idea of the direction of the correlation curve. What such data cannot provide is the correlation itself. To get a suggestion for the shape of the correlation curve, even if it turns out to be linear, we need at least three data points, which calls for testing factors at three levels. To render the correlation more dependable, we need to test factors at more than three levels. But the limitation is the number of experimental runs to be planned and the concomitant resources for the experiment itself, followed by the analysis of the data. Thus, the number of levels for testing factors in most cases will be the "most we can afford," not "all we can do."

The effect of two factors acting together at three levels on a dependent variable is the simplest in this category of factorial design. The number of combinations of factors is given by $l^f = 3^2 = 9$. But, three points, each an outcome (recorded as a number) of the dependent variable, may not be adequate for observing the trend—the shape of the curve—of the relation between the dependent variable and the independent variable(s). Testing each factor at four levels will provide a better definition, which, of course, is made possible at the cost of increasing the number of required runs to $4^2 = 16$. After running the experiment, we will have sixteen numbers to deal with, each recorded as the effect caused by one of the sixteen combinations of the two independent variables. Assuming fictitious values for the dependent variable, we will attempt some possible observations in the following.

9.2 Four-Level Factorial Experiments

We take the two factors as a and b, and designate the levels of

a as a_1, a_2, a_3, a_4

b as b_1, b_2, b_3, b_4

The sixteen combinations can be lined up as follows:

$$\begin{matrix} a_1b_1 & a_2b_1 & a_3b_1 & a_4b_1 \\ a_1b_2 & a_2b_2 & a_3b_2 & a_4b_2 \\ a_1b_3 & a_2b_3 & a_3b_3 & a_4b_3 \\ a_1b_4 & a_2b_4 & a_3b_4 & a_4b_4 \end{matrix}$$

We notice in the above matrix that

1. In the first column, the factor a_1 is common, and the only variables are the levels of b: b_1, b_2, b_3, and b_4.
2. In the first row, the factor b_1 is common and the only variables are the levels of a: a_1, a_2, a_3, and a_4.

What the independent variables a and b actually are and how they are regulated and made effective in the experimental setup are of no concern to us. Similarly, what the dependent variable is, how it is recorded, and in what dimensional units do not concern us. What does concern us is having the response—values of the dependent variable, expressed as numbers—correspond to each of the sixteen combinations of the factor levels and analyzing for interpretation relative to the *main effects* of and interactions, if any, between the two factors.

Table 9.1 lists the corresponding values of the dependent variable, alongside the level combinations of the two factors a and b.

Table 9.1 *Two-Factor Four-Level Experiment with Hypothetical Responses*

a_1b_1	a_2b_1	a_3b_1	a_4b_1	6	9	11	12
a_1b_2	a_2b_2	a_3b_2	a_4b_2	9	12	14	15
a_1b_3	a_2b_3	a_3b_3	a_4b_3	7	10	12	13
a_1b_4	a_2b_4	a_3b_4	a_4b_4	10	13	15	16

Two-Factor, Four-Level Experiment with Hypothetical Responses

In this table, only whole numbers are used for convenience in demonstration, though in actual experiments, such values are very unlikely. Figure 9.1, in which the same values are presented graphically, provides the following observations:

1. The four levels of factor *a* are presented on the horizontal scale, also called the *x*-axis for convenience. The locations of points on the *x*-axis do not represent either the quantities or the units. The distances between points, which are equal in the figure, do not indicate that the levels are at quantitatively equal intervals; they were placed so only for convenience. The quantities and the units thereof depend on the experimenter's discretion. The absence of a factor, for example, may be the first point on the *x*-axis, not at zero. The subsequent points, showing the presence of that factor, do not have to be spaced according to their quantities.
2. The vertical scale (also called the *y*-axis for convenience) is made to show the numerical values of the dependent variable; this is done strictly to scale.
3. The four levels of factor *b* do not have a scale of their own. Instead, they appear as four lines, each consisting of three segments: b_1–b_2, b_2–b_3, and b_3–b_4. Each line looks as if it shows the result of a one-factor-at-a-time experiment with *a* varying and *b* kept constant. This is not the case.

The above three features will be common to all two-factor, four-level experimental designs.

Now for some observations specific to the numbers in Table 9.1, rendered graphically in Figure 9.1.

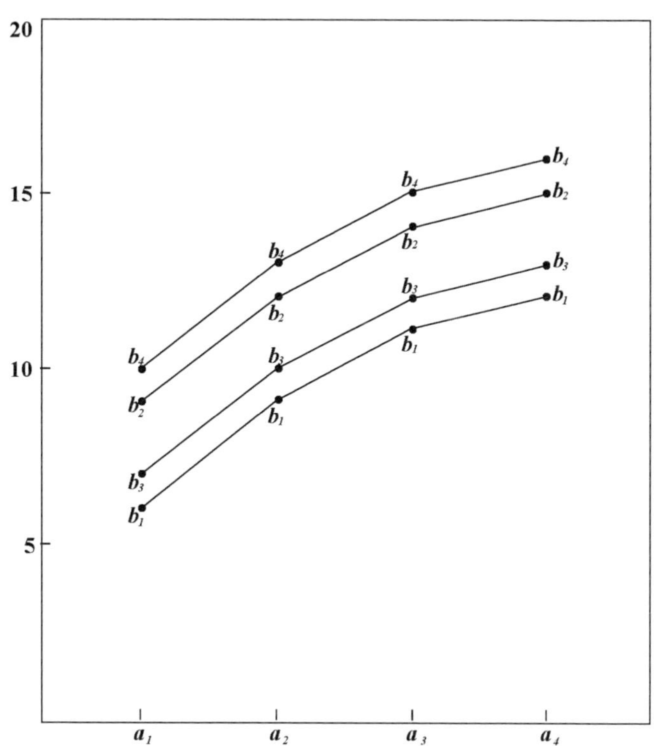

Combinations of two Independent Variables (factors)

Figure 9.1 *Experimental outcome for two factors at four levels (assuming* a *as the prime factor).*

9.2.1 Main Effects and Interactions

Looking for the main effects of factors is quite involved here, unlike in Chapter 8, where there was only one "distance" between the "low" and the "high" levels of each factor. In contrast, here we have three separate distances in series. Referring to the definition of *main effect* in Chapter 8 and applying it to factor *a*, we have three, not one, segments of main effect. These are the effects of changing *a* from level 1 to level 2, from level 2 to level 3, and from level 3 to level 4. Also, with only two levels of factors, it was possible in Chapter 8 to assign a "–" to the low and a "+" to the high levels of factors. Further, taking advantage of the "–1" and "+1" symbolism, analyzing the values of the dependent variable could be done "mechanically" (see Tables 8.8 to 8.11). That advantage is not available for an analysis in which the factors are dealt with at more than two levels.

The four lines of each segment level are parallel to each other. In terms of numbers, we may say the distance between any two levels of factor *b* (which is a measure of the improvement attributable to factor *b*) is the same at different levels of factor *a*. In other words, factor *b* is uniformly effective at different levels of *a*. This is a situation of factor *b* improving (supplementing, we may say) the effect of *a*, but not interacting with it. Thus, the four lines, each corresponding to a level of factor *b*, signal the condition of *no interaction* between the two factors within the range of the factor levels used in the experiment.

As mentioned earlier, it is likely and desirable that either *a* or *b* can be identified as the prime factor. If *a* can be so identified, it is desirable to locate it on the *x*-axis, as is done in Figure 9.1. Suppose in the experiment we are presently discussing that *b* is the prime factor. In such a situation, the same data as we now have in Table 9.1 can be rearranged as shown in Figure 9.2.

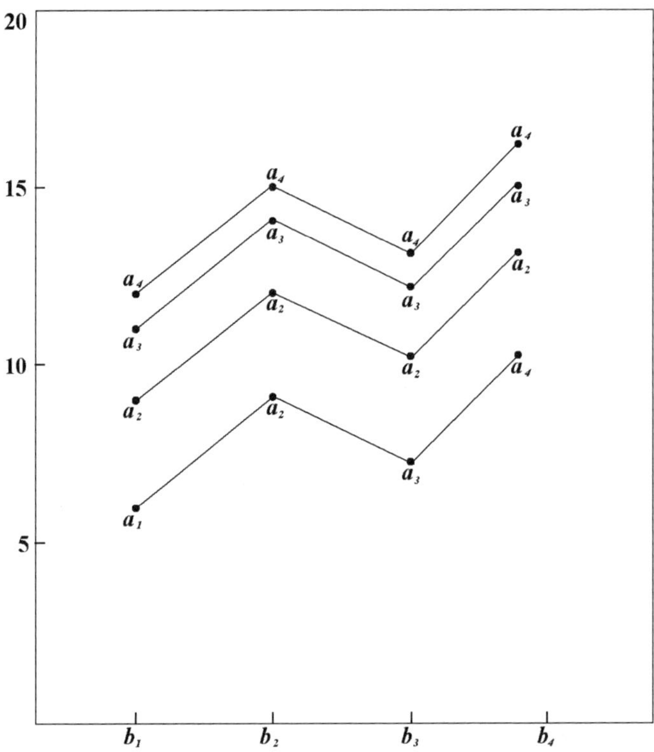

Figure 9.2 *Experimental outcome for two factors at four levels (assuming* b *as the prime factor).*

Comparing this figure with Figure 9.1, we notice that the four lines, each corresponding to a level of *a*, are again parallel to each other, though the shapes of all the lines in Figure 9.2 are different from those in Figure 9.1. We may thus summarize that whatever the shape of the lines corresponding to different levels of the second factor, the lines being parallel is the criterion test for no interaction between the two factors. Now, though it is desirable to identify one of the two factors as the prime factor, it may often be impossible, even presumptuous, to do so. The mere fact of the lines being parallel should be an adequate test for no interaction.

9.3 Interactions

In most cases of research, particularly with quantitative parameters, data as depicted in Figure 9.1, with absolutely parallel lines, are rather rare. A set of data as shown in Figure 9.3, instead, is one of many possibilities. We will examine Figure 9.3 for interaction and for main effects.

We notice in this figure some segments of parallel lines: a_2b_1–a_3b_1 is parallel to a_2b_3–a_3b_3, and a_3b_1–a_4b_1 is parallel to a_3b_2–a_4b_2, but no two lines are parallel through the range a_1 to a_4.

This is a case of widespread interaction. It is possible to make several qualitative statements based on this data; for example:

1. The effect of adding *b*, at level b_2, with *a* is fairly additive.
2. The effect of adding *b*, at level b_3, with *a* is beneficial only up to level a_3.
3. The effect of adding *b*, at level b_4, with *a* is beneficial only at the lowest level of *a*, namely a_1.
4. Any further addition is deleterious.

The above observations are typical of the benefits derived from looking for interactions. Also, it is good to remember that *a* and *b* could be purely or partly qualitative parameters, for example, different varieties of rice treated with different brands of fertilizers or with different quantities of a particular fertilizer. In

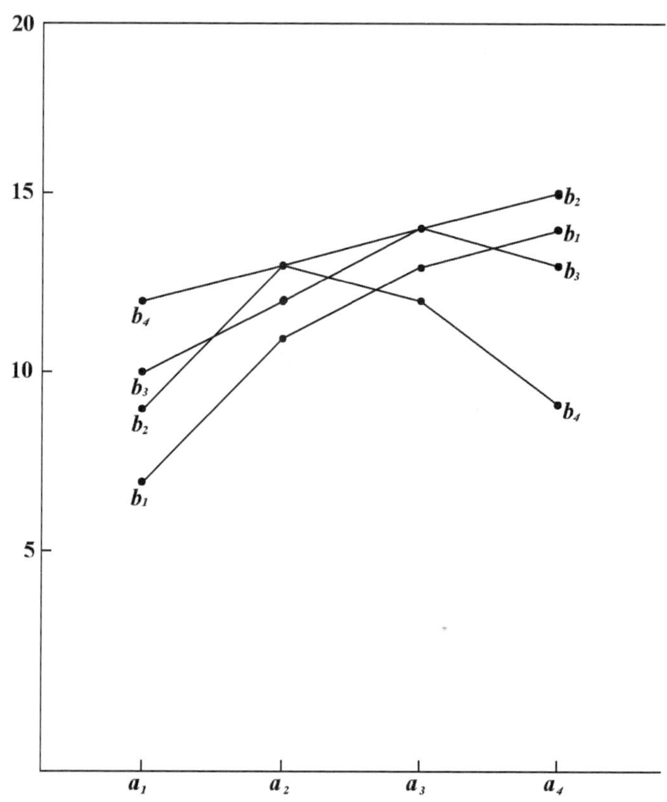

Figure 9.3 *Experimental outcome for combination of two factors at four levels: presence of interactions.*

such cases, in statements such as 1 through 4 above, the names of the appropriate parameter will take the places of the levels of *a* and *b*, and replacing the word "adding" with the phrase "associating (or treating) with" makes better sense.

9.4 Main Effects

For observation of the main effects, we import the data in Table 9.1, augmented with some derived, additional information, now shown in Table 9.2.

We observed in Figure 9.1 that there is practically no interaction between the factors *a* and *b*, but that has no bearing whatever on the presence or absence of the main effects of either *a* or of *b*. Focusing on the first column of numbers in this table, we notice that *a* is tested for its effect four times, each time with a

Table 9.2 *Augmented Data from Table 9.1 for Observation of Main Effects*

Factor b at \ Factor a at	a_1	a_2	a_3	a_4	Responses/Dependent variable				Line average for effects of b
	Independent variables								
b_1	$a_1\ b_1$	$a_2\ b_1$	$a_3\ b_1$	$a_4\ b_1$	6	9	11	12	9.5
b_2	$a_1\ b_2$	$a_2\ b_2$	$a_3\ b_2$	$a_4\ b_2$	9	12	14	15	12.5
b_3	$a_1\ b_3$	$a_2\ b_3$	$a_3\ b_3$	$a_4\ b_3$	7	10	12	13	10.5
b_4	$a_1\ b_4$	$a_2\ b_4$	$a_3\ b_4$	$a_4\ b_4$	10	13	15	16	13.5
					Column averages for Effects of a				
					8.0	11.0	13.0	14.0	

different level of b. The column average 8.0 is thus the average of the effect of a, tested for four times, each time associated with b. Testing the main effects of factors at three or more levels is not as simple as it is for factors at two levels. For instance, the main effect of a has three components: the effect of causing it to jump

1. from level 1 to level 2
2. from level 2 to level 3
3. from level 3 to level 4

These, taken from the table, are respectively

$11.0 - 8.0 = 3.0$

$13.0 - 11.0 = 2.0$

$14.0 - 13.0 = 1.0$

Similarly, the three components of the main effect of b are

$12.5 - 9.5 = 3.0$

$10.5 - 12.5 = -2.0$

$13.5 - 10.5 = 3.0$

We also note that these effects are cumulative in nature. For instance, the main effect of causing *b* to jump from level 1 to level 3 is (12.5 – 9.5) + (10.5 – 12.5) = 1.0, which is the same as (10.5 – 9.5) = 1.0.

Though these numbers give some idea of the combined effects of multiple factors, the values of main effects, per the definition, require further elaboration. Considering their importance, we will take a close-up look at the main effects, using another, even more simplified example. Firstly, we note that the main effect of a factor *a* at level *l*, designated a_l, acting in combination with other factors, by definition, is the difference between the mean of all the values of the dependent variable at a_l and the grand mean of all the values of the dependent variable in the entire factorial experiment. Using only two independent variables, *a* and *b*, both quantitative, acting at only three levels, the main effects, using fictitious numbers, are demonstrated in Table 9.3. Symbols α and β are used for the main effects of factors *a* and *b*, respectively; subscripts with α and β, as with the independent variables, indicate the levels.

Table 9.3 *Evaluation of Main Effects in a Factorial Experiment*

	b_1	b_2	b_3					
	Factor combinations			Average value of the dependent variable			Mean	Main Effects
a_1	$a_1 b_1$	$a_1 b_2$	$a_1 b_3$	6	7	9	7.3	α_1 -1.8
a_2	$a_2 b_1$	$a_2 b_2$	$a_2 b_3$	7	8	12	9.0	α_2 -0.1
a_3	$a_3 b_1$	$a_3 b_2$	$a_3 b_3$	9	10	14	11.0	α_3 +1.9
	Mean			7.3	8.3	11.7	Grand mean 9.1	
	Main Effects			β_1 -1.8	β_2 -0.8	β_3 +2.6		

Denoting Table 9.3 as a (3 × 3) factorial, we may think of other factorials, such as (3 × 2), (2 × 3), (3 × 4), (4 × 2), and so on. The same method as that used in the table of evaluating the main effects for these holds good. Each of the values in the table, indicated as main effects, corresponds to the factor at one level. For practical purposes, though, the so-called differential main effects are more significant; those show the difference in values of the main effects at two different levels of a given factor. Three typical examples are tabulated below.

Differential main effect of factor . . .	between levels . . .			is
a	a_1	and	a_2	–1.8 – (–0.1) = –1.7
a	a_1	and	a_3	–1.8 – (1.9) = –3.7
b	b_2	and	b_3	–0.8 – (2.6) = –3.4

The values of the main effects and the differential main effects being negative need not concern us because the improvement in the dependent variable may be either an increase or decrease in its value based on what it represents.

9.5 More on Interactions

1. Interactions between factors, when there are only two, are easy to identify in two-way tables, as in Table 9.1, or even make visible in figures such as Figures 9.1 to 9.3.
2. The presence of interactions complicates the interpretation of the experimental results relative to the benefits or harmful effects of individual factors.
3. If there are no interactions, each factor may be treated as if it were independent of all other factors. That is when the main effects shine unobscured.
4. In most cases of experimental research, getting data like that in Figure 9.1 is rare. Almost always, some interactions may be expected, but all interactions observed may not be genuine in that some may simply be the effects of uncontrolled or

uncontrollable "noise" factors. To make sure that this is not the case, statistical tests devised for the purpose, known as *significance tests*, need to be applied to the data.

5. Some interactions observed may be such that if the data—the values of dependent variable y—are transformed to other forms, such as y^n (where n is a real number) or log y, the interaction may disappear. It is worth making an effort, using mathematical methods available for the purpose, to get rid of the interactions.

6. In experiments with factors at three or more levels, interactions may exist only in certain segments, between levels of factors. If the factors are partly or fully qualitative, statements in words, such as cautions to be taken, are quite adequate. If, instead, the factors are quantitative, appropriate actions, in terms of controlling the quantities of factors, are necessary.

7. All interactions are not necessarily to be frowned at. Imagine, in a metallurgical experiment to enhance the hardness of steel, that a particular alloying element, A_1, when used alone, is not effective and that another alloying element, A_2, likewise is found to be ineffective. When both A_1 and A_2 are used together, however, the hardness is found to be much enhanced. The interaction in this case is found to be beneficial, prompting further experiments to optimize the percentage additions of A_1 and A_2.

9.6 More Factors at More Than Two Levels

When dealing with more than two factors, each at more than two levels, the significance of main effects and interactions remains unchanged. Now we need to analyze the experimental responses for main effects and interactions with three or more factors. For instance, with three factors instead of the previously analyzed two, the main effects need to be referenced to the specified factor, between specified levels, in the "company of" two other factors.

To that extent, the interpretation of main effects is clear, but interactions of three-or-more-factor, three-level experiments are not that simple. With four factors, for instance, *a*, *b*, *c*, and *d*, the following interactions need to be searched for:

Four two-factor interactions: *ab*, *bc*, *cd*, and *ad*

Three three-factor interactions: *abc*, *bcd*, and *abd*

One four-factor interaction: *abcd*

Analysis of such interactions, limited to two levels of each factor, was done in Chapter 8 (see Table 8.10). There the analysis was rendered simple by the expedient of using "–1" and "+1" for the "low" and "high" levels. As that advantage is unavailable for dealing with factors at three or more levels, the analysis becomes quite complex. Thus, with the factor level jumping from two to three (or more), the analysis is not simply a matter of extension.

The reader is advised to consult a text devoted to the design of experiments for such analysis, when needed. Here, we instead move on to an analysis of experimental responses when there are four factors with each at three levels. This analysis necessarily has to be an incomplete factorial, as was done in Chapter 8 for two levels, but here extended to three levels.

9.6.1 Fractional Factorial with Three-Level Factors

In this chapter so far, we have dealt with full factorials with two factors at three or more levels. With an increased number of factors, the number of runs necessary, if we stay with full factorials, will increase exponentially. When beset with such problems, fractional factorials are the obvious solution. Adapting the example discussed in Chapter 8, we will form a fractional factorial, this time with more specifics and concrete numbers, and strive for the solution.

Example 9.1

We will digress here slightly relative to the selection of factors, based on the input of the Quality Circle and limitations imposed by statistical considerations. It is known that several factors, influencing each other, determine the qualities of manufactured products (or components). To experiment with the influence of all possible factors together is virtually impossible, making it necessary, as a prelude to factorial experiments, to select a few of the many factors involved. If such a selection is made based on the experience of those personnel who are involved with, and know intimately, the various aspects of the product, the benefits of multifactor experiments will be very much enhanced. Following is a brief narration of an example in which an attempt was made to combine the benefits of the concept of Quality Circle with those of factorial experimentation.

This problem, previously mentioned in Chapter 8, is from the metal-casting industry, and the case in point is the need to increase permeability. Less than adequate permeability will result in gas-related defects in castings. Over a certain number of weeks, a greater-than-acceptable number of such defective castings had been found.

In the Quality Circle, it was thought necessary to pick one or, if unavoidable, two factors that have the most influence on permeability. Further, it was the intention to investigate the linear relation of such factors to permeability. The least number of levels that can help us in that direction is three. Testing at three levels involves two components of information for each factor, for instance, the main effect of a given factor between level 1 and level 2 and another between level 2 and level 3. If we choose five factors to start with, as we did in Chapter 8, we end up with ten components of information. The related statistical methods indicated that the least number of runs in a fractional factorial—the smallest orthogonal array—necessary for a three-level factorial is nine. Also the maximum number of information components that can be handled in such an array is one less than the number of runs. That reduces the highest number of factors we can handle and still be able to search for the most influential ones to $(9 - 1) \div 2 = 4$. Knowing these limitations, I posed to the Quality Circle the problem of having to reduce the factors to not more than four. The discussion in the meeting, stated very briefly, proceeded as follows:

1. The sand technician said that angular sands, compared to round sands, expose more surface area,

hence, cut down on permeability. He asked for roundness to be a factor. But round sands are expensive. The foundry had been using what the supplier branded as "subangular" sand. The vice president, who was a member of the Quality Circle and who controlled the inventory for the company, agreed to pay for a sufficient quantity of round sand for the experiment. If and when they decided in favor of round sand for making molds, they would consider the expense.

2. Compression strengths in the test lab ran in the range of 13 to 16 psi. There was no reason to believe that lowering the compressive straight to some extent may considerable increase the casting defects. The sand technician mentioned that higher-than-adequate strengths are the result of excessive compaction and that excessive compaction would lower permeability.

   At this point, the foundry superintendent interjected that while passing through the molding floor, he had observed that most of the molders, even before using the jolt-squeezing machines, use pneumatic rammers to compact the molds. His suggestion to the floor supervisor against the use of rammers was met only with excuses. Now that he was sitting in the Quality Circle, he insisted on taking away the rammers from the molders, but he agreed to wait until the experiment was done. The question was how to test the effect of the degree of compaction on permeability in the lab. The sand technician supplied the answer: instead of the normal three rams on the specimen, he could also test permeability with one and two rams. That became a factor.

3. Two different kinds of sands were supplied to each molding station. The molders were expected to cover the pattern surface with facing sand, about one-half inch thick, by sifting through a riddle—this process is known as *riddling*—then to fill the molds with backing sand. Some molders, instead

of doing that, simply shoveled the facing sand into the mold. The quality inspector suggested that some castings showed rougher surfaces than others, and he attributed that to following or not following the procedures of riddling the sand. He wanted riddling/not-riddling to be one of the factors, but this factor could be made effective only on the molding floor. It was decided that the experiment now in planning should all be done in the lab, not on the molding floor. Thus, riddling was excluded as an unrelated factor, at least for the time being.

4. The sand technician had previously pointed out repeatedly that the sand had excessive fine material. The Methylene Blue clay was in the range of 8 to 10 percent and the AFS clay in the range of 12 to 14 percent. The difference between the two was the measure of dead, fine material (from now on simply referred to as "fines"). The fines filled the pores among grains and worked against permeability. The plant manager, who had to report to the president, had not heeded the sand technician's suggestion to take appropriate actions to reduce the fines in the return sand from the shakeout. Now the sand technician had an opportunity to relate fines with casting rejects. From the small quantity of sand required for the lab tests he could remove the fines by placing the sand on number 200 sieve and shaking; longer periods in sieve shaking would remove more fines from the sand. He could thus obtain several samples of sand, each containing different quantity of fines. He wanted to include percentage of fines in the sand as a factor; it was accepted.

5. Speaking about fines, as this foundry made mostly ductile iron castings, there was a considerable quantity of sea coal in the sand. A nonstandard test known as "Loss on Ignition" could serve as a good measure. Reducing sea coal as much as possible is good for permeability. The metallurgist in

the Quality Circle objected. He argued that sea coal, despite its connection to reduced permeability, had an otherwise important part to play. So, it was dropped from consideration as a factor.

6. Permeability is closely related to the volume of gases generated. Sand has considerable moisture needed to temper the clay; the steam generated is an obvious part of the gases. Reducing moisture is a way of reducing the steam generated; thus, there will be less gas to pass through the bulk of molding sand. The moisture recorded was in the range of 5.0 to 6.5 percent, but this rather high percentage was necessary because of the greater than normal percentage of *live clay*, the clay that is active in forming the bond, also known as Methylene Blue clay. If the moisture were going to be considered as a possible factor, it would make better sense to treat the combination of live clay and moisture together. The sand technician offered to take live clay and moisture together as one factor, mixing them in the same ratio as the current percentages

Table 9.4 *Fractional Factorial of Four Factors*

Factors		Levels 1	2	3	Corresponding designations		
a	Sand angularity	subangular	semiround	round	a_1	a_2	a_3
b	Number of rams to compact permeability specimen	3	2	1	b_1	b_2	b_3
c	Fines passing through #200 screen	~3.5%	~ 2.5%	~ 1.0%	c_1	c_2	c_3
d	Moisture (+ live clay in ratio 6: 9)	>6.0%	5.5%	5.0%	d_1	d_2	d_3

of live clay and moisture; here onwards referred to as moisture (+live clay), it became another factor.

Table 9.4 shows, in summary, the four factors at three levels each forming the fractional factorial for the experiment.

It may be observed that factor *a* is qualitative in that the angularity is not measured; instead, it is accepted with the quality assigned by the supplier. The other three factors are quantitative; the quantities in numbers could be assigned to these, though not very precisely. We would note that the "increments" between levels of these factors are actually reductions because the improvement we expect in permeability is caused by reducing the quantities of these factors. Also, as usual, the distances between levels are not to scale.

After this preparation and background, we are now ready to face, in familiar form, the orthogonal array of four factors in three levels, presented in Table 9.5. The first column shows the serial number of runs, and the last column, the values of permeability (each an average of five tests), corresponding to the factor-level combination on each line in the middle column.

Table 9.5 *Fractional Factorial of Four Factors at Three Levels*

	Independent Variables				**Dependent Variable**
Test run	Factor Levels a	b	c	d	Responses/ Permeability Numbers
1	1	1	1	1	151
2	1	2	2	2	176
3	1	3	3	3	177
4	2	1	2	3	183
5	2	2	3	1	194
6	2	3	1	2	179
7	3	1	3	2	199
8	3	2	1	3	166
9	3	3	2	1	212

With the numbers read off this table, we can now calculate the values needed to know the main effects of all the factors. For instance, such values for factor *c* are

at level 1: (151 + 179 + 166) ÷ 3 = 165

at level 2: (176 + 183 + 212) ÷ 3 = 190

at level 3: (177 + 194 + 199) ÷ 3 = 190

The main effect of *c* between levels 1 and 2 is 190 – 165 = 25, and that between levels 2 and 3 is 190 – 190 = 0. Represented to scale on the *y*-axis (the *x*-axis has no scale), such effects are shown for all the factors in Figure 9.4.

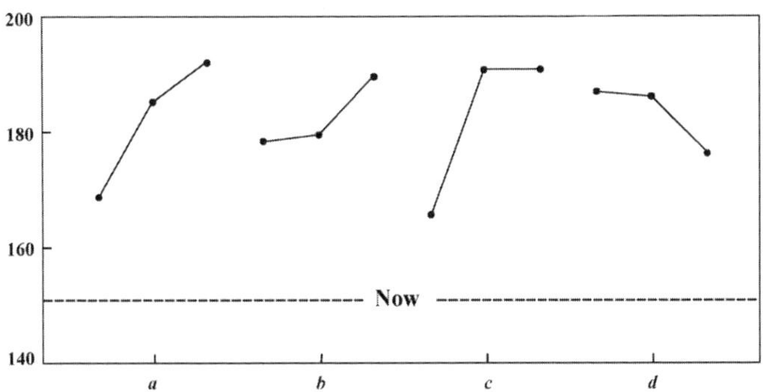

Figure 9.4 *Evaluation of Main Effects in a Factorial Experiment*

The discussion that ensued in the subsequent Quality Circle meeting, guided by this figure, is summarized as follows.

1. Each of the factors tested for was found effective as compared to the "now" average permeability shown in the figure for reference. It was no surprise to anyone; the deliberation done in selecting the factors was complimented.
2. Factor *a*, the roundness of sand grains, as reasoned out by the sand technician, was found to be most effective. It was desirable to work with at least the semiround sand, which was likely to be less expensive than round sand. The vice president in charge of inventory appreciated the prediction of the sand

technician but pointed out that the use of such sand in their system had to remain a wish. To replace the entire *system sand* (the sand that is used repeatedly in the foundry) plus the present stock in the silo with "semiround" sand, would be extremely expensive. Besides, the material being scarce, there were no consistent and dependable sources for sand of that quality. Changing the grain shape of the sand was dropped for the time being as a "not viable" means for the purpose.

3. Factor *b*, reducing the degree of compaction, showed to be a very effective factor. How can compaction of the permeability specimen be simulated in actual mold making? The sand technician was confident that bulk density could be used to relate the two, but the more practical problem was, Will mold compaction equivalent to applying two rams on the specimen, instead of the normal three, be adequate for the mold to withstand erosion by the liquid metal during pouring and solidification? The metal involved was cast iron, which is heavy and very hot. Despite such risk, should they implement the experimental finding? The molding supervisor was hesitant. The foundry superintendent offered a compromise; only for big molds, with flask depths greater than 9 inches, would pneumatic rammers be allowed, not for others. Any problems related to not using rammers would be reviewed promptly.

4. Factor *c*, removal of fines, recommended itself very favorably. The foundry superintendent was very impressed. He thought that this, perhaps, was the best and the quickest change to implement; it could be done entirely within the plant, independent of outside resources, men, or material. He offered to discuss with the company's president the need for, and possibility of, enhancing permeability. He said he would use this opportunity to invite the president and, through him, the plant manager to become part of the Quality Circle.

5. Factor *d*, reducing the bond quantity, sounded very reasonable before the experiment. Now a gain in permeability was recorded even with the current, excessive quantity of bond. Even more puzzling, with decreasing bond quantity, the permeability decreased, instead of increasing. The sand technician was deeply disturbed; should they continue with the present level of bond, even with reduced fines? Others said yes; his concurrence was not forthcoming.

I did not take sides in the dispute. Instead, I made a note that this might be a good case of severe interaction masking the main effect. In the course of time, as the fines progressively decreased, the demand for quantity of moisture (+ live clay) might decrease with it. Meanwhile, to study the interaction, I was thinking of an experiment with two factors: percentage of fines and quantity of moisture (+ live clay), each at four or more levels, keeping the other factors out. I did not mention this at this meeting; it is for the future.

9.7 Bibliography

Diamond, W. J. *Practical Experimental Design for Engineers and Scientists.* New York: John Wiley and Sons, 2001.

Lee, W. *Experimental Design and Analysis.* New York: W. H. Freeman and Co., 1975.

Mead, R. *The Design of Experiments.* Cambridge, U.K.: Cambridge University Press, 1988.

Montgomery, D. C. *Design and Analysis of Experiments.* 5th ed. New York: John Wiley and Sons, 2001.

Part III: The Craft Part of Experimental Research

10

Searching through Published Literature

Knowledge is of two kinds. We know a subject, or we know where we can find it.

—*Dr. Samuel Johnson*

Studying literature is part of doing research. To the researcher, this task is more than digesting prescribed textbooks, as is the case in the course work of most undergraduate and graduate students. Besides naming some of the sources of literature, this chapter points out that, to the researcher, searching literature is more like finding one's way along a trail in a jungle than taking an evening walk through the city park. An example shows that discrimination is an essential ingredient and that doing the most does not always result in getting the best.

10.1 Researcher and Scholar

Research scientists, of necessity, need to become scholars. The best scholars are not necessarily the best researchers, and vice versa. Knowledge in science and technology is so diversified that a professor, an expert in his own area of specialization, is nearly a novice in regard to the knowledge of other professors in the same department. It is not uncommon that dozens of Ph.D. candidates working in the same department know very little of each other's work. Each researcher, in a sense, is at the narrow apex of a small pyramid of specialized knowledge. To a lesser extent, this is also true at the master's level and at the level of the undergraduate thesis, in that order, the pyramid being smaller. To be able to do research, the researcher should have a good idea of the breadth and height of the knowledge pyramid wherein he is placed. But to study every "block and brick" of the pyramid, that is, every published article in his area, is impossible. Fortunately, unlike in the pyramid analogy, not all articles are equally important; some are even insignificant. How well one can identify the significant ones among the hundreds that are available is a matter of gift, luck, and, of course, experience.

For a newcomer, to survey the literature is a formidable task, but much depends on what kind of preparation one needs to begin the task. Suppose, as it was pointed out in Chapter 1, one wants to do research for a Ph.D. and has a modest scholarship, but his superior has suggested or advised no direction, or "track." For him, the task of surveying the literature is very wasteful and frustrating. At the other extreme, one may become, from the beginning, a member of a team working, not necessarily for a degree, under the direction of a guide, manager, or director. Then, to choose literature in one's track is fairly easy.

We will assume a case at neither of these extremes, in which the candidate is not a member of a research team and has the guidance of a knowledgeable superior who provides a track, suggests the context of a problem, but lets the researcher find his own way and define the problem for himself. This is the most desirable case for the researcher. Where should such an independent, rather self-dependent, researcher start his search in literature? We attempt to outline here how he can make progress.

Thanks to computer technology, we live in an age of information explosion. Taking the hint from what the school children do when their teacher gives them a project, we naturally go to the Internet. There is a lot of "information" on any topic, on any meaningful word combination one may form. The list of studies available is simply overwhelming, sometimes running to several hundred items. If the researcher spends the time to read each one of the titles and looks at the source to make a reasonable guess as to its relevance, he may get a few articles he thinks worthwhile to look at in detail.

10.2 Literature in Print

Printed literature, though not available at one's fingertips, is more likely to be fruitful. In the order of building the base and then the height of the pyramid, one may start with the general encyclopedias like the *Britannica* and the *Americana.* Next in order are encyclopedias on individual broad subject areas, such as physics, chemistry, biology, electrical engineering, and materials. Then follow handbooks on various subject fields, wherein a fairly detailed discussion can be found on various topics. Then come books in the field, which are often too numerous, and it is not easy to see all of them. Looking through the contents of selected ones, with luck, the researcher may notice a whole chapter dealing with material that he is looking for. Chapter-end and book-end bibliographies are a great source for more literature. *Books in Print* lists all of the books currently available, including recent publications, classified by subject. Recent developments, maybe in one's area, will likely be dealt with in one or more of these books. Careful perusal of the contents of those that suggest themselves by their title may bear fruit.

Then follow periodicals (e.g., journals, annals, transactions). Most journals occasionally publish volumes devoted to abstracts of articles or indexes of articles published in previous issues of that journal, and in some cases, in other journals as well. Based on the title and abstract description, you may trace or call for the reprints of relevant articles. Occasionally, there are articles wholly devoted to a review (or survey) of available published literature in a particular problem area.

In doing a literature search, there is no orderly, step-by-step procedure to follow. Quite a bit of roundabout search is fairly common. The researcher may find, for example, in a reference in a recent article another article closer to his interest, or a monograph, or even a book, published quite a few years ago dealing with material he has been searching for. To find such a jewel is very gratifying. By the time the researcher reaches this stage in the literature search, he may get the frustrating feeling that whatever he thought of as a problem has already been "solved" by others and that he has appeared on the scene too late.

Not so. In fact, such frustration is an unconscious indicator that the researcher has done a "good enough" amount of searching and that the end of the tunnel is around the corner. Almost always, unconsciously, his mind has been groping toward defining the problem. And almost always, when he has least been thinking about it, he will see the problem defined, delimited, and presented to his conscious mind. That problem then becomes his.

He then knows all the aspects that have been researched already and has found that one aspect that somehow escaped the notice of others and that he is called upon to adopt and make his own. The question or questions that define the problem, when changed from interrogative to assertive form, take a new life in the shape of a hypothesis. And that is what the researcher has been waiting for: a hypothesis. From then on, hypothesis becomes the focal point of his research work.

The literature survey does not end there. In one sense, it begins over again, but with a better focus, at a higher level in the knowledge pyramid. We notice that the literature survey begins with the problem; it started when the problem was only a track, a suggestion, a context. On the other hand, more literature survey was needed to define the problem, to delimit it and to form a hypothesis. How much literature survey is needed to reach that stage is a matter of luck and anybody's guess. And when does the literature survey end? It does not; all through the next steps of experimentation, analysis, writing, and presentation, literature survey is a continuous and integral part of research work. In fact, it almost becomes a habit with research workers; even when they change careers, they tend to remain scholars.

10.3 Overdoing?

It is interesting to ask if there is anything like too much scholarship, meaning too much study of literature. Opinions differ among research workers, but I know at least one case in the affirmative. I recollect a great scholar, let us call him "Rao," from my early research years. Rao was meticulous in everything he did, very hard working, devoted to knowledge, and brilliant as well. There was no book he had not seen in his area of study, no recent or old article that he had not studied. He was sure to be present at all seminars and colloquiums, not only in his department but often in others as well. He was a familiar face at all meetings of scientific interest. He missed no lecture by a guest speaker. One could bet that no presentation was complete without Rao asking questions and that those questions were very pointed and relevant. Often he would answer his own questions because the presenter could not. His comments were almost like the last words on the topic under discussion. Many of us expected great things from Rao. If he were to win a Nobel price, we would not be surprised.

Rao emigrated to the United States, to the disappointment of many because they would miss him, and to the approval of many more because, to them, that was the right atmosphere for Rao's genius to flourish. Rao joined a well-known university in the Midwest and worked under a well-known professor. Several years later, I emigrated to the United States too. Rao was in touch with a common friend and came to visit that friend, who was now married with children, had worked many years in industry, and had returned to another university to earn a Ph.D. We were chatting on many things about our common past: changes in our old institution, our professors and co-students there, who was where, doing what, most of whom we had lost track of.

Rao was now balder than before, with some gray hair here and there, but the same old smiling, casual, unreserved self—and still unmarried. Unexpectedly, almost as a confession, he talked about himself. His words were not exactly these, but their meaning was as follows: "At that time, I was very close to my subject. I still tend to be. I have come to understand one thing when it is too late. Being too close to the subject is not conducive to creative work. I see a lot of details, which is not good. A certain amount of distance between the subject and the observer is essential. I miss that." I felt a great pity, almost sadness for him. He was of course smiling and casual. There was no trace of self-reproach in his statements. I remembered a saying relevant to artists, but I could not say it to him: "When observing the woods, do not look at the trees."

10.4 After the Climb

Returning to literature survey as a part of research work, it is possible, with some overlap, to think of two phases of that: (1) before hypothesis, beginning with the inception of the problem, and (2) after hypothesis, lasting throughout that research and beyond. In the second phase, the researcher looks for a few selected journals relevant to the research, more specific articles in those journals, and even within an article for more specific information. He is now more likely to be rewarded by studying current publications more regularly, more selectively, and more closely because now he is one of those who are extending the knowledge front in that specific field. It is also possible that he knows who the leading authorities are in that field in different parts of the world, who in a sense became his colleagues and coworkers, though at long distance. He may even contact, correspond with, and get to know them personally. Now, that all-too-eager communication technology aids him with e-mail addresses, voice mail, and fax machines!

10.5 Bibliography

Gash, Sarah. *Effective Literature Searching for Research.* 2nd ed. Hampshire, UK: Gower Publications, 2000.

Building the Experimental Setup

When should the construction of an apparatus be abandoned because an apparently better idea has presented itself? It is obvious that if better ideas followed one another fast enough, it could happen that no apparatus would ever be finished.

—*E. Bright Wilson Jr.*

Experimental setups are, more often than not, put together or fabricated by researchers. The combination of a research project and a researcher most often tends to be anything but standard; individual differences prevail. This chapter discusses some of the general principles involved in building experimental setups, such as deliberated appropriateness, simplicity, compactness, and safety. Also, the main events in getting ready for the experiment are discussed, including judicious selection of vendor items and selection of measuring instruments and their calibration. It is pointed out that cost should not be taken for granted and that the researcher's time in building the setup should be considered as a cost item.

11.1 Diversity to Match the Need

Perhaps no other aspect of experimental research, as an organized activity, has so much diversity as the setup, often referred to as the apparatus, used for carrying out the experiment. With some reservation, we may say that there are as many setups as there are researchers. Notable exceptions are the standard equipment used repeatedly, either in succession or during the same period on a time-sharing basis; a good example is an electron microscope, usually one-of-a-kind equipment in moderate-size research labs. But even here, elaborate as it is, it is only a piece of equipment, seldom used "stand alone," unconnected with any other auxiliary equipment. In terms of the "connections," either physical or schematic, each investigator is likely to have his or her own idiosyncrasies. For instance, it is quite reasonable to expect that an experimental setup in a chemistry lab will be different from that used in an electrical engineering lab; the difference relates to pieces of equipment, the nature of connections among them, the measuring instruments for recording the data, the safety devices, and so on. Besides, considering the researcher's need and purpose, further subject to the individual viewpoint and approach, each setup, in a sense, is the contrivance of the particular researcher. Hence, we restrict the discussion here only to those aspects that are somewhat common to most research equipment in diverse subject areas.

11.2 Designing the Apparatus

To design is to contrive for a purpose; purpose is the all-important consideration. The first question any researcher needs to ask, as a way of designing the apparatus is, What do I want to observe? The answer to this question is part of the hypothesis. To the extent that the answer to the above question is explicit, the design is made easy. If, instead, the hypothesis can provide only an implicit answer, the task of the experimenter is to revisit, redefine, and reword, if necessary, the hypothesis to obtain an explicit answer. When the question is explicitly answered, it is a matter of detail to know whether what to be observed is an object, a phenomenon, or a measurable quantity. To suit the nature of the object or the phenomenon, one needs to decide the mode of perception, the one mode most commonly used being sight. Considering the limitations of human perception, for instance, with the "invisible" band of electromagnetic radiations, several devices of detection or amplification are often found necessary. Now, returning to the key role of hypothesis in designing the apparatus, there is reason to expect innumerable variations. For convenience, in principle, we may think of two sides to the apparatus: the *cause* side and the *effect* side, The two can be in both space and time, close or widely separated; each side may consist of one or more stages, and each stage may, in turn, contain many devices, measuring instruments, and connections among those. Considering the various fields of study, various hypotheses, and, above all, various human decisions, it is hardly possible even to hint as to what ought to be the structure of, and how to build, an experimental setup, but there are some common features in the effort of designing the apparatus, of which experimenters can avail themselves.

11.2.1 Seeking Advice

Firstly, it is a desirable trait of the experimenter to be resourceful, professionally inquisitive, and open. This sounds like a truism, but it is not uncommon for an individual researcher to be overly private, even sensitive, about his hypothesis, hence, the kind of setup he may need to experiment with. Projecting his attitude onto other colleagues and coworkers, he may shy away from asking questions relative to the details of a setup they may be using, even in the next room. He may not need the same pieces of equipment, but the logic as to why certain pieces of equipment are chosen, among other possibilities, and why the different pieces are put together in that particular arrangement: that itself is a valuable education. Many researchers reinvent the wheel, though not so dramatically, even those who are close and accessible. Visiting a few researchers in related areas, observing their apparatus, and candidly asking for advice is likely to yield handsome dividends. While one may wish to be original, a rare gift, imitation is the sure way to learn. Many pitfalls may be avoided by gratefully seeking a helping hand.

11.3 Simplicity, Compactness, and Elegance

Guided by the hypothesis, one may visualize the kind of arrangement among several pieces of equipment to constitute the setup. It is worth one's while to review each conceived piece of equipment individually, asking oneself crucial questions: Is it possible to eliminate a given piece of equipment and still accomplish the purpose? Is there a substitute that is less elaborate? Is it possible to combine its function with that of another? And so on. A sketch similar to a block diagram is helpful in visualizing at this stage. One may be surprised to learn that the first impression is not necessarily the best. As in many other aspects of life, a little introspection may reveal the many wasteful means to accomplish the given end. Fewer pieces of equipment require less space and fewer connections, and fewer things can go wrong. It is not rare to find, on close observation, a researcher building a setup with all the care and elaboration of a spider weaving its web, with this difference: the researcher's setup could be revised through much thinking and rethinking before he started to build. Sitting amid a very elaborate web of equipment may impress peers and supervisors, but if it is built in a hurry, instead of serving as a means, the web may become itself an end, the researcher trapped with many avoidable headaches.

Arranging at one end of the room (often a "cage" in a university lab) the different pieces of equipment as close to each other as possible but allowing access to each, having the controls (for example, electric switches) and measuring devices (including meters) arranged on a switchboard, clearly naming and marking each item, and having desk space at or near the entry is a better arrangement than scattering the various items all around and having the desk in the middle. Appropriate grounding of electrical connections, shielding against radiation and heat, and safety devices should be present and double-checked. Ample light at the entry and inside the room, dampening mechanical vibrations, adequate ventilation, and so forth, should be given as much care as the setup itself. The elegance of housekeeping varies among individuals, and it is likely to show in one's setup. But below a level necessary, it is likely to lead not only to clumsiness and confusion, but to danger as well.

11.4 Measuring Instruments

There are very few experiments in science and technology, even when only an object or a phenomenon is to be observed, in which no measuring instruments are involved. And when the change in a measurable quantity constitutes the effect (sometimes referred to as *outcome* or *quality characteristic*) part of the hypothesis, the need for measuring instruments is all too obvious. The extent and kind of instruments needed, besides the elementary ones for the measurement of lengths, weights, volumes, time, temperature, and so forth, which are taken for granted as available to all researchers, depend on the field of study and the accuracy required, as dictated by the hypothesis. A few terms and concepts, which apply to most instruments, follow.

Instruments may be broadly divided in a somewhat overlapping fashion into two classes: gages and meters.

Gages are usually replicas of the corresponding shape and dimensions to be measured, used on a comparative basis as references, for example, the "Go and Not Go" gage for checking the tolerance of a hole. But we also speak of "pressure gages," where pressure is directly read on the instrument; in strict terminology, it ought to be a "pressure meter." With similar freedom, we also refer to the "depth gage," "strain gage," "radius gage," and so on. The word "gage" has traditionally been used for other purposes also as indicating thickness of sheet materials, the diameters of wires, the width between railroad rails, the size of gun barrels, to name a few.

Meters, in conservative terminology, are instruments that read in real time on a cumulative basis the quantity involved (e.g., water meter) or the intensity of a property over a wide range (e.g., thermometer). But the word "gage" has been widely used, often subsuming the meaning of "meter."

Resolution is the smallest difference in dimension that an instrument can present. For instance, a ruler with a millimeter division has a resolution of 1 mm. On the other hand, a micrometer (without a vernier) has a resolution of 0.1 mm. *Sensitivity* is an alternate term used.

Precision refers to the reproducibility or repeatability of the same reading at different times and in different environments; for example, a steel ruler is likely to be more precise for measuring lengths than a wooden ruler, considering that wood expands with high humidity and contracts with high temperature. Often, wrongly, the term *accuracy* is used in place of precision.

Accuracy refers to the extent of closeness of a measurement to its "true" value, as judged by an appropriate standard. For instance, a ruler measurement of, say, 2.31 cm may be compared with 2.31 cm obtained by stacking gage blocks on a surface plate. To the extent that the ruler reading coincides with that obtained with gage blocks, the ruler is accurate.

All the above references were made, for convenience, to length measurements, other length-related measures being thickness, diameter, taper, flatness, tolerances, and so forth. All of these are dealt with in an area of study known as metrology. The terms described above also apply to measurements of volume, weight, and time and, further, even to measuring instruments such as pressure gages, ammeters, and voltmeters.

In the course of their application, hence, at the point of selection, the measuring instruments are to be judged relative to a few more considerations:

- *Speed of response* indicates how rapidly the instrument indicates the measurement. This is of particular importance when many readings are to be taken at short intervals.
- *Stability* is the virtue by which an instrument holds its reading over an extended duration (without fluctuations).
- *Linearity* is referred to as being "high" when an instrument presents accurate readings over its full working range.

11.5 Calibration

For most of the primary and secondary measures, there are measuring instruments, like rulers for lengths, balances for weights, and thermometers for temperatures. In daily routines, we take such measuring instruments for granted. If a person's temperature is noticed to be 104°F on a newly bought thermometer, one does not react with complacence, wondering, How am I sure that the thermometer is not wrong and that the actual temperature is four degrees less? An experimenter in science, on the other hand, relative to a thermometer involved in his experiments, needs to wonder exactly that. He cannot act on faith and let things be as they are. The chance that a thermometer in his experiment is incorrect may be only one in a thousand, but it is necessary that he knows if it is reading the correct temperature, and if not, how far it is off. The actions taken as a way of answering these questions, plus the corrective measures implemented if needed, are known as *calibration.*

No measuring instrument, however brand-new or expensive or fancy looking, is above the reach of doubt leading to calibration. And here is a task that demands scruples and conscience on the part of the experimenter, not to be delegated to someone else lightly, for the simple reason that if the experimenter is beginning with less than dependable measurements, he cannot make dependable correlations. If, after making some progress, he is required to repeat the experiments, now with better readings, the fatigue is self-inflicted besides the waste of time.

As to the means and methods of doing calibration, however, no general rules can be laid out. Depending on the instrument, the accuracy required, and the reference and standards available, a wide range of possibilities and "acceptability" prevail. Most of the basic standards, for example, the "correct" length of a meter, are supposed to be maintained in all countries by an organization like the U.S. Bureau of Standards. Copies of such standards are likely to be available in many significant reference places, such as the metrology lab of a local college or university or research organization. But every experimenter who needs to measure a length need not run to a metrology lab. More often than not, a copy of a copy of . . . the locally accessible standard is adequate. This is where the question of the "needed accuracy" arises. An investigator, heat-treating a weld specimen, for example, does not have to measure the soaking period to the accuracy of seconds; minutes are good enough. Here again, the experimenter needs to decide how much accuracy is required for each of the measurements involved. A sense of proportion, not to be taken for granted, is an asset in an experimenter. Fortunately, there are statistical methods to help determine such issues when in doubt. But in most cases, comparing with one or two near-by standards may be adequate. For example, a dial indicator used in an experiment can be calibrated using the gage blocks with surface plate within the same lab. All that is needed is to check and correct, if need be, the measuring instruments in use beyond any reasonable doubt.

There is another kind of calibration we have not dealt with, which is, indeed, more often referred to, without qualification, as *calibration*. The issue we have dealt with so far is then considered an issue of "standards." When a quality to be measured is such that a one-to-one correspondence exists, or can be reasonably established, with another quality that is more easily or more precisely measurable, then the instrument used to measure the second quality is said to be "calibrated" in terms of the first quality. Often, reading electrical units, like current and voltage, is more convenient than reading some physical property, say temperature. For example, if we are required to measure, over a range, the temperature of molten glass, we cannot think of using a thermometer with a glass stem and mercury bulb. Instead we can use a thermocouple made of a "standard" pair of metal wires, which can be dipped safely in the liquid glass bath. The (emf) voltage generated in this circumstance, using that particular pair of wires, has a one-to-one correspondence with the temperature of the bath, this information having been well established in the body of science and available in many standard handbooks, which we normally need not cast doubt on. Then, we record the voltage values corresponding to the variation of temperature taken by adding heat energy to, or removing it from, the glass bath. (It is herein assumed that the voltmeter was independently subject to calibration relative to voltage readings.) Using a correspondence relation, either in graphical or tabular form, for every voltage value, we can find a temperature value. That is, for every reading on the voltmeter, there prevails a corresponding temperature of the bath. The voltmeter, at this level of use, is said to be "calibrated for reading temperature." This indeed is the principle used in all commercially available temperature recorders.

11.6 Researcher as Handyman

The circumstances in which an experimenter finds himself determine the variety and degree of skill demanded of him. To researchers in humanities, mathematics, and "theoretical" sciences, this concern for building the experimental setup is just hearsay. At the other extreme, almost all areas of engineering and technology require, for experimental research, some kind of setup to be designed and put together. Between these extremes, most sciences require apparatus of varying complexity. Some fortunate ones have simply to "slip in" to the existing equipment, which has been used in previous years. On the other hand, some researchers may find themselves in places and among people where research has no tradition. Starting from scratch to build research equipment is a formidable task. Here, brainwork has to take a back seat, at least until tangible, usable equipment is available. In the process of getting that accomplished, it happens more often than not that the researcher has to do one or more of the jobs that carpenters, machinists, electricians, plumbers, and welders, if available, could do better. It is not a coincidence that, the presence of lab technicians in most university labs, graduate students are often busy in workshops instead of in labs. The technicians often do not, or do not want to, understand the specificity of the project and the equipment needed; they conveniently let the students swim or sink in their own mess. Needless to say, those who have the skills of the handyman have the advantage of overcoming hurdles in the way of building experimental setups. It is prudent of the researcher, as a part of his training, to acquire the basic skills of many trades, even if he does not enjoy those handyman activities. And it would be thoughtful of the universities and other research labs to require that their future researchers be initiated in basic skills.

11.7 Cost Considerations

The cost of an experimental setup may be considered against two items: (1) the cost of the various pieces of equipment, and (2) the cost of the experimenter's time. In places where research is part of the tradition, several items, like motors, gear trains, pulleys, tubes, and clamps, may be available, waiting to be poached from a previously used, now junked, setup. It is quite exhilarating to those who enjoy such work to pick up pieces here and there and to put them together to form a purposeful whole. It is indeed a desirable trait for experimenters not to dislike junk.

Though it is tempting to use such items, the time it takes the researcher to extricate, recondition, and match this junk needs to be carefully considered; it may often turn out to be a very expensive exhilaration. If such junk is not adequate or is not available, the choice is to obtain all or some of the items, new or reconditioned, as vendor items. Items less expensive but adequate for the purpose are preferable to those of the best quality demanding a premium price. The cost of items paid for is obvious and hence usually gets the required attention, whereas the cost of the researcher's time escapes unnoticed. The time spent in putting the pieces of equipment together no doubt gives the familiarity needed in using the setup, but it is the kind of work that others, whose time is less valuable, can do. That the time spent by most researchers is not counted by punching time cards is a privilege, which, to remain so, should be used for justifiable purposes.

Quite often a researcher inherits the setup, his work being to continue, or extend with changes, the research that was being done by his predecessors; this is often the situation of graduate researchers in universities. The cost accountable to setup in such cases is almost nil. The researcher has reason to congratulate himself that he can straight away move on without having to build the bridge, namely, the setup. Like many other phases of experimental research, at the time of planning to build the experimental setup, there are likely to be many unknown factors, at least in detail. The experimenter should deliberate on all the factors known at the time. Among the many paths that suggest themselves, the one to be followed should incur the least amount of time as well as money. Research, however exciting the activity may be, still needs to be justified on the mundane grounds of weighing needs and benefits on one side and costs of all kinds on the other. Quite often, this decision is not easy, especially if the experimenter is not experienced. This is where the help and advice of experienced peers, seniors, and superiors (in the hierarchy) is indispensable because it is not a purely "research" decision but a "management" decision, tempered with research experience, that needs to be made. One of the important questions is whether to build in-house or to buy the equipment if it is available on the market.

Though each person's experimental setup is unique, it is advisable to use as many components as possible as vendor items. As in the case of requiring a burette in a chemistry setup or a gear in an engineering setup, the vendor market offers a lot of items at relatively low cost. Even bigger units, which have several components already assembled, are available on the market at competitive prices. Awareness of the availability of manufactured items should be a part of the resourcefulness of the experimental investigator, or at least, of his advisors and seniors. In almost all cases, the manufactured items cost a lot less than the sum of the costs of the corresponding components, and the assembly is more optimal, and the performance more assured, than the corresponding built-from-scratch devices. An amateur radio operator may build his own radio for a hobby, no matter how much it costs and how long it takes. That hobbies cost a lot more time and money than we can justify on a purely materialistic basis is well known. To mix a hobby with building an experimental setup can, more often than not, be counterproductive.

Should an amateur electronics hobbyist be allowed to build a ten-point temperature recorder as a project in an industrial research and development (R&D) department? I witnessed a case in which it was allowed. Let us call the amateur Pete, and his boss at that time Jamy. Jamy was the manager of R&D with nothing to commend him to that position except his being a very inoffensive yes-man to his boss, the senior manager. When I came on the scene, Pete was very seriously fabricating something, the expected outcome of which was practically unknown to his colleagues. Jamy had great admiration for Pete's skill, and Pete had been working on nothing but this mysterious equipment for the previous six months. Jamy promptly approved all the cash purchases—every other week, a few hundred dollars—for components for the equipment. Only after a whole three months and after pointed questions did I learn that Pete was building a temperature recorder. After more than ten months of devoted work, Pete finally declared that the setup was finished.

I was required to test-perform the temperature recorder in a real-life customer problem, in which temperatures at different points of the system were significantly related. I traveled with Pete and another technician to the customer's place, over seven hundred miles away, driving a station wagon full of equipment, fairly bulky and heavy, with quite a few loose pieces to be assembled on the spot. We stayed in a motel and worked two weekend days in the plant, installing Pete's equipment to have it ready to be tested on Monday when work started. Finally, when we began seeing the readings, we found the readings to be erratic, not reproducible, and totally incoherent. Needless to say, Pete, after completing the fabrication, had not done any calibration before this use.

Having previously used a twenty-four-point temperature recorder, a product manufactured by Leeds and Northrup, I could see the colossal amount of time and money wasted in the name of R&D. To say anything uncomplimentary to Pete was uncalled for, and I faced a dilemma: whether to betray Pete or to betray the company.

Fortunately, Pete left the company for "better pay" two months or so after this incident. Nobody ever touched his contrivance after that, and like many other inspired follies, it gathered dust in a corner. If the manager had been reasonably resourceful and had any experience with research, he could have purchased a ten-point temperature recorder as a manufactured item at a cost less than one tenth of that incurred by Pete, and ten months earlier. The situation here is that the manager of R&D failed to distinguish between ends and means, and Pete found and exploited the manager's ignorance to promote his hobby at the cost of R&D resources.

11.8 Bibliography

Wilson, B. E., Jr. *An Introduction to Scientific Research.* Mineola, NY: Dover Publications, 1990.

Part IV: The Art of Reasoning in Scientific Research

12

Logic and Scientific Research

For Aristotle, logic was the science of proof or evidence, and it should be the same for us today.

—D. S. Robinson

The process of scientific research is guided by the principles of logic, both inductive and deductive. Since logic is taught as a liberal arts discipline in colleges, students of science tend to bypass it in their course of study. Thus, an opportunity is lost to learn the language of science, which is necessary in doing the groundwork for research. Some of the more important terms of logic, as applied to scientific research, are reviewed in this chapter.

12.1 The Subject, Logic

Philosophy is a typical subject of liberal arts. Logic is a part of philosophy; either at undergraduate or graduate level, it is taught by professors in the Department of Philosophy, unless there is a Department of Logic at the particular college. For the purpose of delimiting the scope of this book, we may act as if philosophy were unrelated to experimental research, but there is a contradiction in this. Logic, which is an essential aspect of experimental research, is the medium used in philosophizing, and few persons planning research careers in science and technology take logic as a subject of study; this is unfortunate.

When I was a college student, like many others, I did not know that research work involved logic. When thought was required for research, either in planning an experiment or analyzing the results, I relied on common sense, often backed by the experience of my peers and advisors. Now I know that it was an unfortunate situation that logic was not a required course and that I was not told by my advisors, who themselves had not studied logic, about its benefits. I also know now that the language of scientific research, as well as the process of reasoning, is similar to the language of logic. My research would have been easier, and more fruitful and elegant, if I had the help that could come from the study of logic. I strongly recommend to people planning careers in scientific research, either experimental or theoretical, to take a course or two in logic as a part of their preparation. There are many excellent texts in logic and advanced logic for self-study or for those who missed the chance in college. The most that can be done in this book is to introduce in this chapter a few terms that the reader may encounter in his or her research career, terms that are common currency in logic. Later chapters will introduce the modes and methods of reasoning relative to analysis of experimental results.

Even before approaching the terms, let us first look at some of the definitions of the subject "logic"; three taken at random follow:

1. "The science whose chief end is to ascertain the principles on which all valid reasoning depends, and which may be applied as tests of the legitimacy of every conclusion that is drawn from premises."[1]
2. "Logic is the study of methods and principles used to distinguish good (correct) from bad (incorrect) reasoning. . . . This definition must not be taken to imply that only the student of logic can reason well or correctly. . . . But given the same native intelligence, a person who has studied logic is more likely to reason correctly than one who has never thought about the general principles involved in that activity."[2]
3. "The laws of logic supply the skeleton or framework within which the test of coherence applies. . . Coherence cannot be accepted as giving the *meaning* of truth though it is often the most important *test* of truth after a certain amount of truth has been known."[3]

These definitions may be incomplete and not fully satisfactory, but together, they provide the idea of what the subject of logic is about. Needless to say, a career in science involves the reasoning of others and often the reasoning of oneself, hence, the help that one can get from the study of logic.

12.2 Some Terms in Logic

With the definition and the context of logic in scientific research now understood, let us look at some terms used in logic and their context in reasoning:

1. *Definition*: a group of words that assigns a meaning to some word or group of words. Every definition consists of two parts: (a) the word or group of words that is supposed to be defined, known as the *definiendum*, and (b) the word or group of words that does the defining, known as the *definiens*. For example, in the sentence "Elephant means a large, herbivorous, gray-colored mammal on four legs," "elephant" is the *definiendum*, and all other words following the word "means" together constitute the *definiens*.
2. *Term*: any word or arrangement of words that serve as the subject of a statement. Terms may be proper names, such as Washington, South American, or Somerset Maugham; common names, such as bridge, behavior, city; or descriptive phrases, such as "pages of this book," "those who can vote," "winner of Nobel Peace Prize." Words or combinations of words that cannot serve as terms include verbs (e.g., *sings*), adverbs (e.g., *courageously*), adjectives (e.g., *beautiful*), prepositions (e.g., *beyond*), conjunctions (e.g., *and*), and nonsystematic arrangements of words (e.g., *signature like camel if*).
3. *Propose*: to offer for consideration, discussion, acceptance, or adoption.
4. *Proposition*: that which is offered (as above).

5. *Premise*: a proposition laid down as a basis of argument.

   Example: All snakes are reptiles

   Curly is a snake.

   Therefore, *Curly* is a reptile.

   In the above argument, the statements of the first two lines are premises, and the statement in the third line is the conclusion.

6. *Conclusion*: the statement in an argument that follows from the evidence presented in the premises (see [5] above).

   Conclusions appear in literature dealing with research in several forms, beginning with, to name a few,

   "It follows from . . ."

   "It is evident . . . "

   "One may observe . . ."

   "Therefore . . ."

   "Based on…"

   "Hence . . ."

7. *Argument*: a group of statements, one of which, the conclusion, is intended to follow from the other statement or statements, the premises.

   It should be noted that there is similarity between the arguments used in the way of forming "conclusions" from "experimental results" by scientists and the arguments presented by attorneys in courts of law.

8. *Evidence*: that which enables the mind to see the truth. It may be from perception by senses, from the testimony of others, or from induction of reason.

   In the context of experimental results, perception is mostly by eyesight in the form of observing something happening or reading a measurement. Hearing, smelling, tasting, and touching have yet to become common modes of observation in scientific research. Relative to testimony of others, all research is, in a sense, an addition to, or modification of, the existing body of knowledge in a particular area or areas of science. To the extent that a particular researcher's findings or arguments substantiate or conform to the existing body of knowledge, that finding or argument is considered to be sound evidence. The inductive argument (yet to be defined and described here) based on reason is, as mentioned above, a finding derived not from observation alone but from a chain of reasoning (conclusions of one argument may serve as the premise for another argument, and so on) until the final conclusion is proposed to be in conformity with the existing and accepted body of knowledge.

9. *Implicate* (*the act of implication*): to show or prove to be connected or concerned (with something).

   There may be a statement in scientific arguments beginning with, for instance,

   "It can be implied . . ."

   "The implication of this . . ."

10. *Inference*: the process of reasoning, whereby starting from one or more propositions accepted as true, the mind passes to another proposition or propositions, whose truth is believed to be involved in the truth of the former. For example, in scientific writings, one may see statements like

    "It may be inferred . . ."

    "The inference from this is . . ."

    "This leads to the inference . . ."

11. *Truth*: conformity to facts or reality, as of statements to facts, of words to thought, of motives (or actions) to profession.

    The first part of this definition, namely, conformity to facts or reality, is especially significant to scientific research in that the effort of science, in the broadest sense, is to add to or correct our knowledge relative to facts or reality. The next part of the definition, namely, statement of facts, becomes significant in a research report, a conference presentation, a thesis, or another form of publication.

Incidentally, "truth" is a much broader term than implied above. Another bunch of dictionary meanings of "truth" includes honesty, sincerity, virtue, and uprightness. Though these qualities are also required in the pursuit of science, their relevance is more "societal" in the broadest sense.

I remember an incident in childhood of getting fact and truth confused. I, like many other children, had been told that to speak the truth is a virtue; the more truth, the more virtue. Truth in my mind became equivalent to fact. So, I and a bunch of children gathered together, sat on a mat in front of the house from where we could watch the slow traffic in the road, and competed in the serious game of gathering virtue. We tried to outdo each other by narrating facts like

There is a man with a turban.

A girl is walking behind him.

There is a donkey carrying a load.

I see a dog behind the donkey.

There is a stick in the hand of the man.

I see a woman far away.

There is a bundle on the head of the woman.

I see a boy pulling a goat.

He has tied a rope to the goat.

Whoever uttered a fact first was entitled to the virtue assigned to that part of the truth. People got virtue points proportional to the number of such statements made.

12. *Reality*: actual being or existence of anything in distinction from mere appearance.

Simple though it seems, philosophers have long wrestled with distinguishing reality from appearance. Several "isms," such as idealism, realism, empiricism, and existentialism, all center on the concept of reality. As is well known, there is no "proof" in philosophy; hence, one may say without embarrassment that all the efforts relative to separating reality and appearance are so far unsuccessful and quite likely to remain so. And what is significant is that, fortunately for the experimental scientist, in most of his routine undertakings, such a tangle is uncalled for.

13. *Inductive arguments* (*induction*): not, as it is supposed to be, an argument that proceeds from particular to the general, but an argument in which the conclusion is intended (by its author) to follow only probably from the premises.

Example:

Premises:

Russell came from New Bedford High. He was a good student.

Goethe came from New Bedford High. He was a good student.

Monet came from New Bedford High. He was a good student.

Conclusion: All students coming from New Bedford High are good students.

But Fermi, who is coming this year from New Bedford High, may turn out to be a dud.

Inductive arguments are very common in experimental research. We do not, and often cannot, exhaust listing all members of a class. After testing "sufficient" (but limited) members of a class, we feel emboldened to make a general statement about all members of that class. With a limited number of experiments, we claim to have found the truth. For example, consider a simple experiment testing the effect of an independent variable, x, on a dependent variable, y. We take several values of x within the required range and find by experimentation the corresponding values of y. Then, we plot a graphical relation between x and y. A straight-line relation of the form

$$y = ax + p$$

is the easiest possible. Otherwise, we attempt a general relation of the form

$$y = ax^n + p, \; (n \neq 0)$$

for a curved line. Relative to the number of (x, y) combinations, we decide, on an arbitrary basis, what is sufficient. However large this sufficient number may be, we are far from exhausting the possibilities. Such decisions are very common in experimental research, particularly in technology, where our inquiry is directed more toward knowing the *empirical*, meaning "likely," relation, rather than the *truth* of the relation.

14. *Deductive arguments* (*deduction*): not, as it is supposed to be, an argument that proceeds from the general to the particular, but an argument in which the conclusion is intended to follow *necessarily* from the premises.

Example:

Premises:

All engineers are intelligent.

Chrysler is an engineer.

Conclusion: Chrysler is intelligent.

Though less often than inductive arguments, deductive arguments are often used to validate experimental results.

12.3 Induction versus Deduction

Some overlapping, hence confusion, between induction and deduction is possible because it is left to the author of the argument whether to stress the probability, even when the probability is very high, or simply to equate very high probability with certainty. Once an elderly colleague of mine took a few days off in a hurry to go for a thorough physical checkup. His son was a medical intern who possibly advised him. I do not know if there was any preceding anxiety, but his sudden disappearance caused some anxiety among his colleagues. Pleased to see him after a few days, I asked, "So, what did they say?" His answer: "They said, I am going to die." Being slightly upset, I asked, "What do you mean?" He answered with a little chuckle, "They said, I am surely going to die. Only they don't know when or how." Now, catching on to his dry humor, I laughed with him. And then, he told me that everything was found to be "normal," as is the case with most physicals. Case in point: Death is certain for all humans. He is a human, so he will die sooner or later. This is deduction. But how do we know that death is certain? The only evidence we have is that we do not know of anybody surviving, say, after age 200; the oldest recorded human was 122 at death. But we have not exhausted the testing of all the members of the class. In fact, we cannot, because there is no tangible proof that, in the distant future, humans will not evolve such that, maybe, one among them will survive forever. Surely none of us believes this, but the issue is not belief; it is proof. In this way, the conclusion is based on a "probable fact," not just a fact; hence, the conclusion itself may turn out to be only a probability, meaning, my colleague may, probably, live forever (God bless him)! This is induction.

1. *Experiment:* a situation that is deliberately set up by an individual or a group of investigators, either for the purpose of demonstrating or finding an object or phenomenon as expected, for verifying a hypothesis, or for validating a theory.
2. *Empirical*: that part of the method of science in which the reference to actuality or factual finding allows a hypothesis to be formed, which, on further *verification*, may become a general principle.
3. *Verification*: the demonstration of the correctness of a hypothesis by means of experiment or observation.
4. *Hypothesis*: a provisional assumption made as to how events—things or phenomena—are interrelated, which is used as a guiding norm in devising experiments and making observations until it is verified or disproved by experimental evidence.

   Broadly we may think of two levels of hypotheses:

   a. *Preliminary hypothesis*, wherein only the facts so far known are used for building the logical structure, a theory, of interrelations

   b. *Final hypothesis*, or "crucial test," which is deducible from the theory, leading to new facts that can be experimentally verified. Observation of such facts directly or of the consequences implied by such facts serves as confirmation of the theory itself.
5. *Theory*: a coherent network of facts and principles. Whereas a hypothesis is a proposition, yet to be tested with evidence for its truth-value, a theory is based on inferences drawn from principles that have been established with independent evidence.

A clear demarcation between hypotheses and theory is not practical in that a hypothesis that has passed some tests of experimental evidence is often claimed to have acquired the status of a theory, though it may still have to undergo one or more crucial tests. A theory of high authority and accuracy is often referred to as a *law*.

6. *Thesis*: an essay or dissertation upon a specific or definite theme, as an essay presented by a candidate for a diploma or degree. The oral examination preceding the confirmation of the diploma or degree is often referred to as a "defense." In this context, a thesis is understood to be a proposition that the candidate advances and offers to defend by argument against objections.

12.4 References

1. *Webster's Universal Dictionary of the English Language*, vol 2 (New York, NY: The World Syndicate Publishing Co., 1937) 976.
2. I. M. Copi and C. Cohen, *Introduction to Logic*, 8th ed. (Macmillan Publishing Co., 1990), 3.
3. Bertrand Russell, *The Problems of Philosophy* (Oxford University Press, 1948), 123.

12.5 Bibliography

Campbell, N. *What Is Science?* Mineola, NY: Dover Publications, 1953.

Cohen, M. R. *A Preface to Logic*. Mineola, NY: Dover Publications, 1977.

Copi, I. M., and C. Cohen. *Introduction to Logic*. 8th ed. New York: Macmillan Publishing Co., 1990.

Johnson, W. E. *The Logical Foundations of Science*. Mineola, NY: Dover Publications, 1964.

Meyerson, E. *Identity and Reality*. Mineola, NY: Dover Publications.

6. *Thesis*: an essay or dissertation upon a specific or definite theme, as an essay presented by a candidate for a diploma or degree. The oral examination preceding the confirmation of the diploma or degree is often referred to as a "defense." In this context, a thesis is understood to be a proposition that the candidate advances and offers to defend by argument against objections.

12.4 References

1. *Webster's Universal Dictionary of the English Language*, vol 2 (New York, NY: The World Syndicate Publishing Co., 1937) 976.
2. I. M. Copi and C. Cohen, *Introduction to Logic*, 8th ed. (Macmillan Publishing Co., 1990), 3.
3. Bertrand Russell, *The Problems of Philosophy* (Oxford University Press, 1948), 123.

12.5 Bibliography

Campbell, N. *What Is Science?* Mineola, NY: Dover Publications, 1953.

Cohen, M. R. *A Preface to Logic*. Mineola, NY: Dover Publications, 1977.

Copi, I. M., and C. Cohen. *Introduction to Logic.* 8th ed. New York: Macmillan Publishing Co., 1990.

Johnson, W. E. *The Logical Foundations of Science.* Mineola, NY: Dover Publications, 1964.

Meyerson, E. *Identity and Reality.* Mineola, NY: Dover Publications.

13

Inferential Logic for Experimental Research

The central problem of logic is the classification of arguments, so that all those that are bad are thrown into one division, and those which are good into another.

—*Charles Sanders Pierce*

Experiments lead to inferences, and inferences lead to conclusions. Words are the medium through which we move from experiments to inferences to conclusions. The appropriate use of words is, thus, the key to deriving the right inferences from the experiments. This chapter begins with well-recognized pitfalls in the use of words, known as *logical fallacies*. The mental process of deriving conclusions is the product of argumentation done within oneself. Formal statements, known in logic as *propositions* or *premises*, serve as the vehicles for the thought process. The varieties of the movements of these vehicles, as formulated by logicians, are summarized in this chapter. These "movements" are explained in two steps: (1) categorical propositions, and (2) categorical syllogisms. A discriminating walk through the pathways clearly charted by logicians is a fruitful exercise for the experimenter. He will find himself better equipped to derive the right inferences, hence conclusions, from his experiments.

13.1 Inferential Logic and Experimental Research

That logic is the art (or science?) of reasoning is a truism. The purpose of any research, particularly experimental research (so far as it can be distinguished from theoretical research), is to draw inferences from the results, as recorded or observed, of the experiments. The inferences can be in the form of either statements or arguments. Should an experimental researcher, to be able to draw proper inferences, be a student of logic? As pointed out in a Chapter 12, it is desirable, though not absolutely necessary. "Logic gives us norms for recognizing correct or good thinking, as well as incorrect or bad thinking, and develops in us a habit of analyzing our thought, of distinguishing carefully between our evidence and our conclusion, and of adverting to the structure of our arguments. . . . [I]t helps us know for certain whether or not our evidence justifies our conclusion."[1] An experimental researcher, whose function it is to derive conclusions from evidence, or his experimental results, cannot afford to ignore the benefits that can be derived from the study of logic. Having said this, should one go a step further and conclude that to be able to do experimental research, one should have completed a few courses of logic for academic credit? No. Researchers at the graduate level and professional researchers may get the reading material from a large number of logic texts available. Undergraduates, who anticipate doing experimental work as a part of their capstone theses, may well be advised to take a course in logic, if one is available in time. If one is not available or convenient, students may be instructed in, or advised to study, a few selected chapters from an undergraduate-level text. Developing proficiency in various aspects of logic is difficult and time-consuming, besides being not essential in the preparation of an experimenter. The one aspect of logic of immediate benefit to the experimenter, and briefly discussed here, is that dealing with the validation of arguments leading to scientific inferences. The material presented in this chapter may serve the reader as a foundation to build on, if and when it is found necessary.

13.2 Logical Fallacies

Inasmuch as a given researcher, more often than not, bases his own research on the published research of others in the field, he needs to understand the way inferences are drawn. And when it is his turn to publish his own research, he should make every effort to make his inferences easily understood. In either event, he should watch out for some pitfalls in thought, as well as in written or spoken language, that have been identified by logicians and are referred to as *fallacies*. Following is a brief review of the important ones.

13.2.1 Fallacies of Ambiguity

As with *definitions* (dealt with earlier in Chapter 2), ambiguity can mislead the reader or listener, often unknowingly and sometimes intentionally. A few variations follow:

1. *Equivocation* refers to using the same word (or phrase) within a statement with two (or more) different contextual meanings.

   Example: Waiting for *leaves* to *fall*, I have hard work waiting this *fall*, which I expect *leaves* me tired.

2. *Amphibole* is the name given to the fallacy where the meaning of a statement is rendered indeterminate because of the loose or awkward way of combining the words.

   Example 1: The dog fought hard to save the puppy, but it died.

   Example 2: The political party in this election that is very corrupt, is likely to be chosen. (Which is corrupt, the political party or the election?)

3. The *fallacy of accent* is committed when, depending on which word (or group of words) is accented, we are led to different meanings.

   Example: In summer, storing chemicals in the attic is not a good idea. (Try accenting "in summer," "storing," and "chemicals," alone, each time.)

4. The *fallacy of composition* is said to have been committed when the properties of the parts are taken obviously to manifest as properties of the whole.

   Example: Cotton is expected to be a "light" material, but a mattress made of cotton can be heavy.

5. The *fallacy of division* is the reverse of the above, wherein the property of the whole is taken obviously to manifest as a property of the parts.

   Example: Americans are tall as a group, but there are some who are short.

13.2.2 Fallacies of Irrelevance

Relevance in either drawing inferences or advancing arguments based on given premises is not as obvious in practical affairs as we normally assume. In research, because of the nature of the discipline, irrelevance is found less than in practical affairs, but deliberate effort is still necessary to keep it in check. Some major fallacies that arise because of irrelevant connections between premises and conclusions are reviewed below. Latin names, which are traditionally attached to fallacies, are also mentioned in parentheses.

1. *Appeal to inappropriate authority (ad verecundium)*: Making reference to opinions or judgments of authorities (or experts) within a given field of study is common, and indeed helpful. Such reference should be confined to the particular field of study. An expert in an unrelated field of study should not be cited as an authority. For example, a worker in chemistry appealing to the opinion of Bernard Shaw or a researcher in tennis quoting Einstein is inappropriate.
2. *Argument against the person (ad hominem):* When, in the process of disputing (or disagreeing) with an inference or a judgment, the author is attacked with the purpose of discrediting his work, the fallacy is known as *ad hominem*, meaning "against the person." In courts of law, political campaigns, and religious propaganda, the *ad hominem* fallacy is generously used, linked with such criteria as "credibility," "honesty" and "trustworthiness." There should be no room for such in scientific research.

13.2.3 Other Miscellaneous Fallacies

1. *Argument based on ignorance (ad ignorantiam):* When a proposition is claimed to be true because it has not been proved false, or when a proposition is claimed to be false because it has not been proved true, this fallacy is committed. A pure believer and an uncompromising agnostic, arguing on whether God exists or not, may both use this fallacy to score some points. But in anything claiming to be scientific, its use is undesirable.

2. *Begging the question (petitio principii):* Also known as circular reasoning, this fallacy is committed when we have assumed what we set out to prove.

   Example: I want to prove that the Ten Commandments, as given in the Bible, are the words of God.

   A skeptic's questions: "How do you know?"

   My answer: "It says so in the Bible."

   A less serious version of this fallacy is our assumption that the laws of nature will operate in the future as they operate at present, which is known as the *principle of induction*. Philosopher David Hume takes strong exception to this, pointing out that using the fact that we have seen something happen during the past up to the present as proof that it will happen in the future amounts to a circular argument. We can hope that the principle of induction in nature may prevail, but we cannot prove it.

3. *Fallacies of accident and reverse accident:* These have particular significance in experimental research because both are related to generalization, and both serve as reminders that generalizations can be hasty and that they may, and quite often do, have exceptions. Witnessing widespread affluence in the United States, a casual visitor may be led to the inference that there are no "homeless people" there, even though we know there are homeless

people, some of whom may starve or freeze to death in winter. This is an example of applying generalization to particular cases; it is known as the *fallacy of accident*. Similarly, witnessing one or two slums in Mumbai, a foreigner might think that all people in India live in wretched conditions. This is an example of carelessly generalizing from a particular case and is known as the *fallacy of reverse accident*.

4. *Fallacies of appealing to emotion (ad populum), pity (ad misericordium), and force (ad baculum):* An orator's speech to fellow countrymen at times of war usually contains several examples of *ad populum*. The language of the defense attorney pleading for lenience in sentencing a criminal may contain many *ad misericordiam* "arguments." And the government of practically every country, pursuing the principle of "Might is Right," engages in *ad baculum*, which may find expression in the language of political leaders. Fortunately for science, these fallacies pose no problems because they are little used.

5. *Fallacy of complex questions:* This refers to asking a question in such a way that the conclusion, often an accusation, is hidden in the question.

   Example: A father to his teenage son: "Have you stopped drinking yet?"

   Again, fortunately, no experimental researcher is likely to ask, or to be asked, such questions.

13.3 Argument

Argument is the intellectual process in which one or more *premises* are used to arrive at a conclusion, which is, in fact, the inference of the argument. A premise is a statement asserting or denying, but not questioning, something, the truth-value of which is taken for granted. When a conclusion can be reached with only one premise, the inference is said to be *immediate*, and the form of argument is known in logic as a *categorical proposi-*

tion. When two premises are involved, with the second premise serving as a medium, leading to the conclusion, the inference is said to be *mediate*; such an argument is known in logic as a *categorical syllogism*. The two forms of arguments are separately dealt with below.

13.4 Categorical Propositions

Propositions are statements. *Categorical propositions* are statements about categories, that is, about groups or classes. A *class* is a collection of all objects (or entities) that have some special characteristics in common. Categorical propositions are used both as premises and conclusions in the process of *deductive argumentation*, in which the premises are expected to provide adequate grounds for the truth of the argument's conclusion. The Theory of Deduction, attributed to Aristotle (384–332 BC), deals with the relation between the premises and conclusions of deductive arguments and further helps determine whether an argument is *valid* or not. In a valid argument, if the premises are true, the conclusion of the argument is necessarily true. If this is not so, the argument is *invalid*.

13.4.1 Forms of Categorical Propositions

There are four standard forms of categorical propositions; each is briefly described, starting with an illustrative example:

1. Universal affirmative

   Example: All factories are companies.

   This is an affirmative proposition about two categories: *all factories* and *all companies*. Further, it infers that every member of the first-mentioned class (known as the *subject term*, *S*) is also a member of the second-mentioned class (known as the *predicate term*, *P*).

   Formula: All *S* are *P*

2. Universal negative

   Example: No factories are companies.

This is a negative statement and infers that, universally, factories are not companies.

Formula: No *S* are *P*

3. Particular affirmative

   Example: Some factories are companies.

   Not all, but some particular, members of the class of factories (at least one) are also members of the class of companies.

   Formula: Some *S* are *P*

4. Particular negative

   Example: Some factories are not companies.

   Formula: Some *S* are not *P*

13.5 Conventions, Symbolism, and Relations among Categorical Propositions

The four forms of categorical propositions (1) through (4) mentioned above are named, and often referred to, respectively, as *A*, *E*, *I*, and *O* propositions.

The (verb) terms of affirmation (e.g., "is," "are," and "were") and negation (e.g., "is not," "are not," and "was not") are referred to as indicating the *quality* of affirmation or negation of a given proposition.

The (adjective) terms ("all," "some," and "no") are referred to as indicating the *quantity* in a given proposition; "all" and "no" are *universal* (meaning complete) in quantity, and "some" is *particular* in quantity.

The *cupola* refers to the term between and connecting the subject term and predicate term in a given proposition, such as "is" in "All S *is* P," or any other word or words (together), usually a verb variation.

Following the conventions relative to terms, the format common to each of the four propositions, relative to the sequence of terms, can be generalized as follows:

Quantifier, subject term, cupola, predicate term

The term *distribution*, in relation to propositions, is used in a very special sense, far removed from its literal one. A given proposition is said to *distribute* the subject term, the predicate term, or both if the reference made by the proposition applies to all members of either class. For instance, in the *E* proposition "No *S* are *P*," reference is made to all members of the class *S*, with no exception; hence, this proposition is said to distribute its subject term. And, reference is made to the class *P*, again as a whole, meaning none of the members of *P* is included in the class *S*; hence, the proposition is said to distribute its predicate term as well. In contrast, if any exception is inferred in the class of either *S* or *P*, then that term is said to be *undistributed.*

Based on such analysis, Table 13.1 summarizes the four forms of categorical propositions.

Table 13.1 *Summary of the Four Forms of Categorical Propositions*

Proposition	Name	Formula	Subject Term	Predicate Term
Universal affirmative	A	All *S* are *P*	Distributed	Undistributed
Universal negative	E	No *S* are *P*	Distributed	Distributed
Particular affirmative	I	Some S are *P*	Undistributed	Undistributed
Particular negative	O	Some *S* are not *P*	Undistributed	Undistributed

13.5.1 Opposition

Categorical propositions having the same subject and predicate terms, but differing in quality, quantity (defined earlier), or both, present different truth relations; the differences are collectively given the traditional name *oppositions*, some variations of which follow.

Two propositions are said to be *contradictories* if one is the denial or negation of the other, meaning that they both cannot be true and both cannot be false.

Example 1:

All wrestlers are athletes.

Some wrestlers are athletes.

Example 2:

No wrestlers are runners.

Some wrestlers are runners.

Example 1 has *A* and *O* propositions, and Example 2 has *E* and *I* propositions. Whereas the propositions of Example 1 differ both in quality and quantity, those of Example 2 differ only in quantity. Both exemplify contradictories.

Two propositions are said to be *contraries* if the truth of one entails the falsity of the other, although both can be false (together).

Example:

All wrestlers are giants.

No wrestlers are giants.

Two propositions are said to be *subcontraries* if they cannot both be false, although both can be true.

Example:

Some tennis champions are tall (people).

Some tennis champions are not tall (people).

The opposition between a universal proposition and the corresponding particular proposition is named as *subalternation*, between which the universal proposition is known as *superaltern* and the particular one as the *subaltern*.

Example 1:

All living things are animals.

Some living things are animals.

Example 2:

All living things are not animals.

Some living things are not animals.

These various kinds of opposition, traditionally diagrammed in a so-called Square of Opposition (named after George Boole), are presented in Figure 13.1.

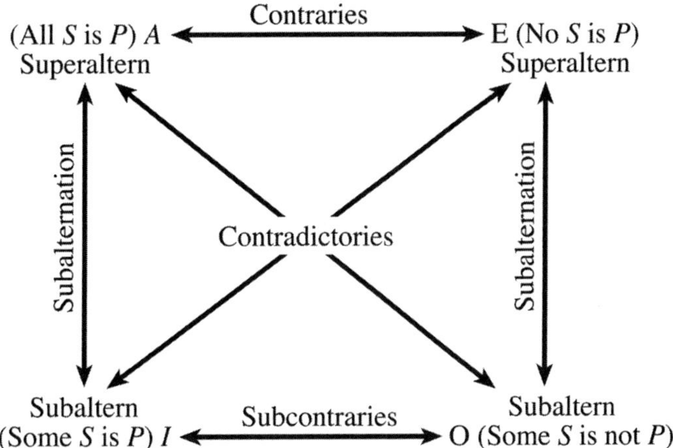

Figure 13.1 *Square of Opposition.*

This can be put to use to draw many, though not all, immediate inferences, as if mechanically; for example, if *A* is true, *O* is false; if *I* is true, *E* is false.

This and many other inferences that can be drawn from the Square of Opposition are listed below:

- *A* being given as true: *E* is false, *I* is true, *O* is false.
- *E* being given as true: *A* is false, *I* is false, *O* is true.
- *I* being given as true: *E* is false, while *A* and *O* are undetermined.
- *O* being given as true: *A* is false, while *E* and *I* are undetermined.

- *A* being given as false: *O* is true, while *E* and *I* are undetermined.
- *E* being given as false: *I* is true, while *A* and *O* are undetermined.
- *I* being given as false: *A* is true, *E* is false, *O* is true.
- *O* being given as false: *A* is true, *E* is false, *I* is true.

13.5.2 More Sources of Immediate Inferences

An application limited to *E* and *I* propositions is shown in Table 13.2. By simply interchanging the places of subject and predicate terms, the truth-value of the converted proposition is retained. This application is called conversion, and each proposition is then referred to as the *converse* of the other.

Table 13.2 *Application of conversion.*

Examples	Proposition	Form
No cats are dogs	E	No *S* are *P*
No dogs are cats	E	No *P* are *S*
Some tennis players are golfers	I	Some *S* are *P*
Some golfers are tennis players	I	Some *P* are *S*

Obversion

Applicable to all the four standard forms of propositions, the process of *obversion* yields a proposition of equivalent truth-value. The process is more involved than *conversion*. Firstly, we need to note the meaning of *complement* as used here. Every category, that is, every class, has a *complementary class*, or simply a *complement*, which is the collection (or class) of all things that do not belong to the original class.

Example: The complement of (the class of) "animal" is "nonanimal," and the complement of (the class of) "nonanimal" is "non-non-animal," which is (the class of) "animal." *Obverting* a given proposition is done by

1. Keeping the subject term unchanged
2. Keeping the quantity of the proposition unchanged
3. Changing the quality of the proposition
4. Replacing the predicate term by its complement

Examples: A few typical propositions to be obverted, known as the *obvertend,* and the corresponding propositions, after doing the obversion, known as *obverse* are shown in Table 13.3.

Table 13.3 *Application of Obversion.*

Example	Proposition	Form	Obvertend/Obverse
All lions are cats	A	All S are P	Obvertend
No lions are noncats	E	No S are Non-P	Obverse
No wolves are nondogs	E	No S are P	Obvertend
All wolves are dogs	A	All S are non-P	Obverse
Some citizens are voters	I	Some S are P	Obvertend
Some citizens are nonvoters	O	Some S are not non-P	Obverse
Some students are not smokers	O	Some S are not P	Obvertend
Some students are nonsmokers	I	Some S are non-P	Obverse

13.6 Diagrammatic Representation of Categorical Propositions

Propositions otherwise expressed in words can also be picturized, using circles to represent classes; some of the symbolic representations in traditional logic and their literal meanings are shown in Figure 13.2.

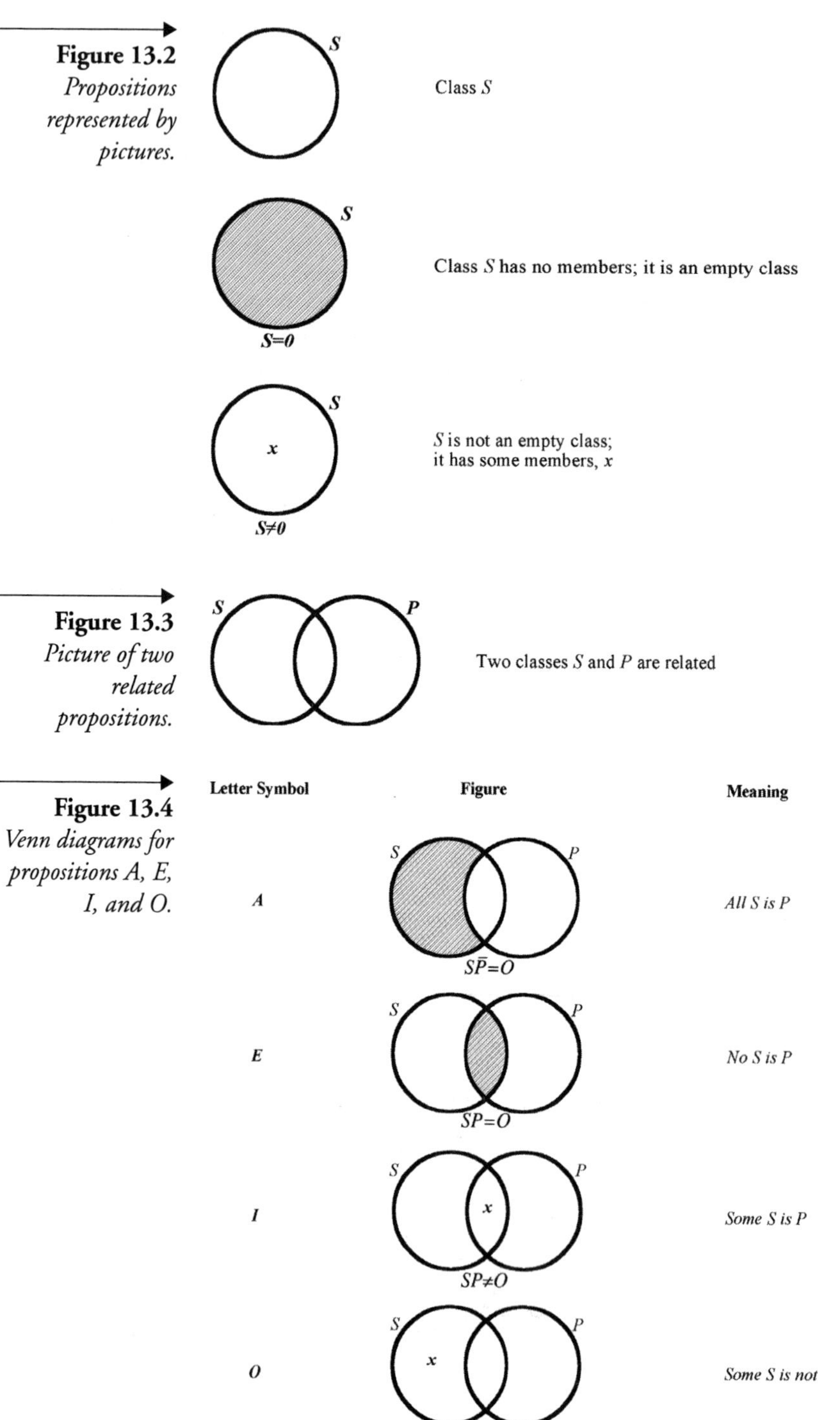

Figure 13.2 *Propositions represented by pictures.*

Figure 13.3 *Picture of two related propositions.*

Figure 13.4 *Venn diagrams for propositions A, E, I, and O.*

A categorical proposition, having to refer to two classes, one each for the subject term and the predicate term, requires two circles for symbolic representation (see Figure 13.3).

Based on these symbolisms, the four standard categorical propositions, each traditionally symbolized by one of the four capitol letters, *A*, *E*, *I*, *O*, can be represented as shown in Figure 13.4. Named after the English logician John Venn (1834–1923), these and similar representation are known as *Venn diagrams*.

13.7 Categorical Syllogisms

A *categorical syllogism* is a deductive argument. As mentioned earlier, it consists of three categorical propositions, the first two of these, in sequence, being *premises* and the third, the *conclusion*. The term that occurs as the predicate of the conclusion is called the *major term*; the premise that contains the major term is known as the *major premise*. The term that occurs as the subject of the conclusion is called the *minor term*; the premise that contains the minor term is known as the *minor premise*. For instance, the following is a typical categorical syllogism:

All glasses are ceramic (first premise)

Pyrex is a glass (second premise)

∴ Pyrex is a ceramic (conclusion)

Here, "ceramic" is the major term; "All glasses are ceramics" is the major premise; "Pyrex" is the minor term; and "Pyrex is a glass" is the minor premise. The third term of the syllogism, "glass," which does not occur in the conclusion, is referred to as the *middle term*. Each of the propositions that constitute a given categorical syllogism is in any one of the four standard forms: *A*, *E*, *I*, and *O*.

13.7.1 Structures of Syllogisms

To be able to describe completely the structure (not the objects of the subject and predicate terms) of a given syllogism, we need to know its two descriptive components: (1) mood, and (2) figure.

The *mood* of a syllogism is determined by the forms of the three prepositions, expressed respectively in the order: the major premise, the minor premise, and the conclusion.

Example:

No saints are landowners: *E*

Some gurus are landowners: *I*

∴ Some gurus are not saints: *O*

The mood of the above syllogism is *EIO*.

Knowing that there are only four forms of standard syllogisms, how many moods are possible? "The first premise may have any of the forms *A, E, I, O*; for each of these four possibilities, the second premise may also have any of the four forms; and for each of these sixteen possibilities, the conclusion may have any one of the four forms; which yields sixty-four moods."[2]

The descriptive component, the *figure*, on the other hand, makes reference to the two position of the middle term within a given syllogism. Symbolizing the major term by *P*, the minor term by *S*, we may think of the following four arrangements. In each the first line is the major premise, the second line is the minor premise and the third the conclusion. In each arrangement, each a variety of syllogisms, the positions of the two M's determine the *figure* of that syllogism, referred to with an ordinal number as shown below.

M—P	P—M	M—P	P—M
S—M	*S—M*	*M—S*	*M—S*
∴ *S—P*	∴ *S—P*	∴ *S—P*	∴ *S—P*
First Figure	Second Figure	Third Figure	Fourth Figure

As indicated above, the four arrangements of the placement of the middle terms are respectively known as the first, second, third, and fourth figures. The syllogism used in the example

above, it may be noted, is one of the second figure. Combining the mood and the figure, the *structure* of the above syllogism can be described as *EIO-2*. With sixty-four variations for mood, each with four variations for figure, 256 (64 × 4) distinct structures are possible, out of which, there are only twenty-four instruments of valid argument.

13.7.2 Validity of Syllogisms

The structure of a given syllogism determines the validity of the argument. And, the validity of an argument is independent of the content or subject matter therein, meaning, a valid argument does not necessarily yield a true conclusion.

For example, the following syllogism is a valid argument with form *AAA-1*, but its conclusion is wrong (that is, untrue), originating from the first false premise:

All animals are meat eaters: *A*

Cows are animals: *A*

∴ Cows are meat eaters: *A*

This serves as an example for the principle that a wrong premise yields a wrong conclusion. If the premises are true, the conclusion from the above form of syllogism will come out true. For instance, the following syllogisms are both of the same form as before and yield true conclusions because each premise in both syllogisms is true.

All metals are electrical conductors.

Iron is a metal.

∴ Iron is an electrical conductor.

and

All fungi are plants.

Mushrooms are fungi.

∴ Mushrooms are plants.

These second and third instances of syllogisms of the form *AAA-1* show a way of testing for the validity of a given syllogism.

Generalizing, we may state that a valid syllogism is valid by virtue of its structure alone. This means that if a given syllogism is valid, any other syllogism of the same structure will be valid also. Extending this, we may also say that if a given syllogism is invalid (or fallacious), any other syllogism of the same structure will be invalid (or fallacious) as well.

Applying this in practice amounts to this: any fallacious argument can be proved to be so by constructing another argument having exactly the same structure with premises that are known to be true and observing that the conclusion is known to be false. This situation generates an uneasy feeling that there ought to be a more formal and surer way to test the validity of syllogistic arguments. One such way is the pictorial (hence, formal) presentation of syllogisms, known as *Venn diagrams.*

13.7.3 Venn Diagrams for Testing Syllogisms

Categorical propositions having two terms, a subject and a predicate, require, as we have seen, two overlapping circles to symbolize their relation as Venn diagrams. Categorical syllogisms, on the other hand, having three terms, a subject, predicate, and middle term, require three overlapping circles to symbolize their relation as Venn diagrams. Conventionally, the subject and the predicate terms are represented by two overlapping circles in that order on the same line, and the middle term is represented by a third circle below and overlapping the other two. This arrangement divides the plane into eight areas as shown in Figure 13.5.

Table 13.4 shows the interpretation of these areas.

These class combinations light up with meanings when we assign *S*, *P*, and *M* to concrete things. For instance, suppose

- *S* is the class of all students.
- *P* is the class of all students studying physics.
- *M* is the class of all students studying mathematics.

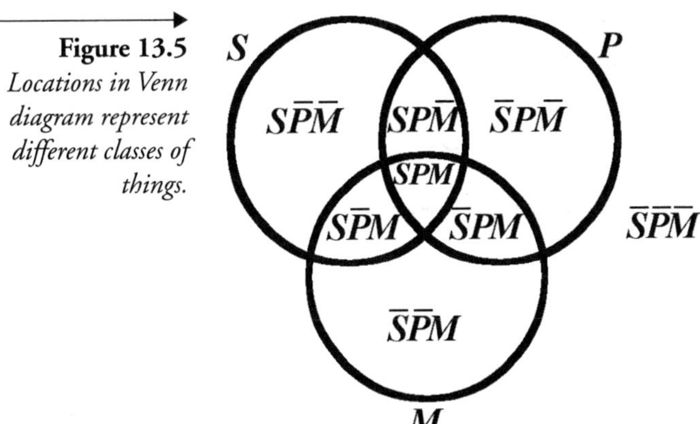

Figure 13.5 *Locations in Venn diagram represent different classes of things.*

Table 13.4 *Description of classes in Figure 13.5.*

Area	Product of . . .	Class of things that . . .
SPM	*S*, *P*, and *M*	belong to all three classes: *S*, *P*, *M*
$SP\bar{M}$	*S*, *P*, and complement of *M*	belong to classes *S* and *P*, but do not belong to *M*
$S\bar{P}M$	*S*, *M*, and complement of *P*	belong to classes *S* and *M*, but do not belong to class *P*
$\bar{S}PM$	*P*, *M*, and complement of *S*	belong to classes *P* and *M*, but do not belong to class *S*
$S\bar{P}\bar{M}$	*S* and complement of *P* and complement of *M*	belong to class *S*, but do not belong to classes *P* and *M*
$\bar{S}P\bar{M}$	*P* and complement of *S* and complement of *M*	belong to class *P*, but do not belong to classes *S* and *M*
$\bar{S}\bar{P}M$	*M* and complement of *S* and complement of *P*	belong to class *M*, but do not belong to classes *S* and *P*
$\bar{S}\bar{P}\bar{M}$	Complement of *S*, complement of *P*, and complement of *M*	do not belong to classes *S*, *P*, or *M*

Then, *SPM* is the class of students who study both physics and mathematics; $S\bar{P}M$ is the class of students who study mathematics but not physics; $S\bar{P}\bar{M}$ is the class of students who study neither physics nor mathematics; and so on.

Further, among *S*, *P*, and *M*, it is possible to construct syllogisms of various moods and figures. For instance,

Some artists are arrogant	*I*	*M—P*
All artists are egoists	*A*	*M—S*
∴ Some egoists are arrogant	*I*	∴ *S—P*

The given syllogism is of the structure *IAI-3*. It now remains to show how a Venn diagram helps us to determine if the above syllogism is or is not valid. The required procedure consists, in sequence, of the following five steps:

1. Identify the names of the *S*, *P*, and *M* terms of the given syllogism.
2. Draw the three-circle Venn diagram (as explained earlier), and label each circle with the right name.
3. Check if one of the premises, major or minor, is universal, and the other particular.

4a. If the answer to (3) is yes, interpret the universal premise first, using two of the three circles of the Venn diagram as required. Then, superimpose the interpretation of the particular premise, using the required two circles for the purpose.

4b. If both the premises are universal, perform the procedure in (4a), with this difference: interpret the first universal premise first, and then superimpose the interpretation of the second universal premise (in the order stated) on the Venn diagram.

5. Inspect the Venn diagram in terms of the conventional zone markings that resulted from the interpretation of the two premises, now having used all the three circles therein. There is no need to diagram the conclusion. Instead, check if the interpretation of the conclusion, in terms of zone markings, is now contained in the Venn diagram.

If it is, the categorical syllogism, as stated, is valid; if not, it is invalid.

Conforming to the above five numbered steps, we will interpret with Venn diagrams, two categorical propositions.

Example 1:

Categorical proposition stated above.

1. *S*: Egoists

   P: Arrogant (persons)

   M: Artists

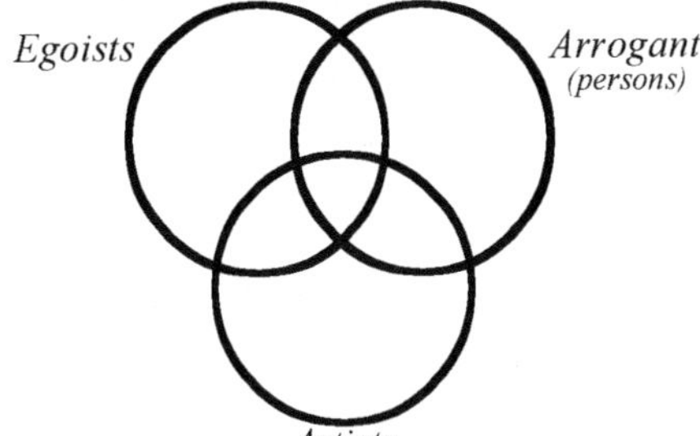

Figure 13.6 *Venn diagram to represent the syllogism.*

2. Figure 13.6 is the three circle Venn diagram representing the syllogism.
3. Only the second premise is universal.

4a. The circle representing the second premise is shaded as in Figure 13.7.

5. Interpreting the conclusion requires that the area of intersection of the two circles "Egoists" and "Arrogant (persons)" contain some members. The zone on the right side of line marked "x" is such an area. Hence, the categorical syllogism, which has the form *IAI*-3, as stated, is valid, and valid are all categorical proposition of that form.

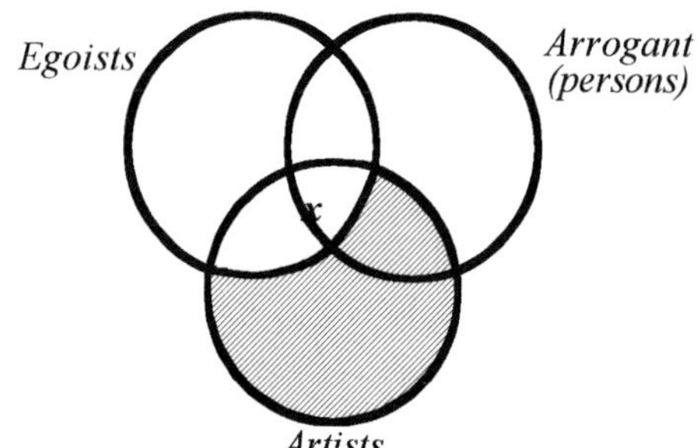

Figure 13.7 *Interpreting the second premise.*

Example 2:

No professors are athletes:	*E*	*P—M*
All athletes are fun-mongers:	*A*	*M—S*
∴ No fun-mongers are professors:	*E*	*S—P*

1. *S:* Fun-mongers

   P: Professors

   M: Athletes

2. Figure 13.8 is the three circle Venn diagram representing the syllogism of Example 2.

Figure 13.8 *Venn diagram representing syllogism of Example 2.*

3. The first and second premises are both universal.

4. Performing the procedure of step 4G results in the Venn diagram shaded as in Figure 13.9.

5. Interpreting the conclusion, as stated, requires that the area of intersection of the two circles " Fun-mongers" and " Professors" should be shaded (cross-hatched). What we have, instead, is clear

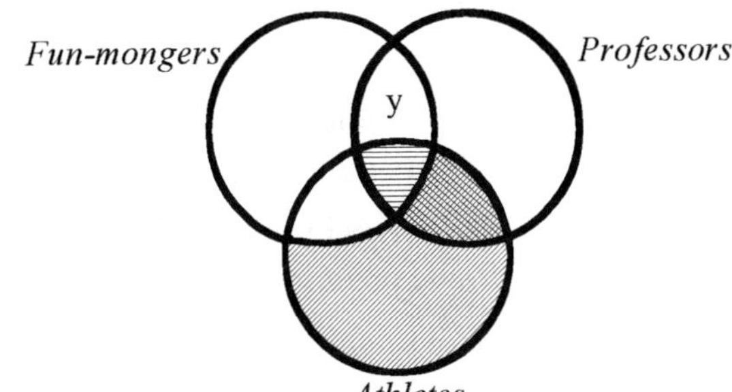

Figure 13.9 *Interpreting the two universal premises.*

area, marked *y*. Hence, the categorical proposition, which has the structure *EAE-4*, is invalid, as are all categorical syllogisms of that structure.

Now, not incidentally, we have also used another criterion, besides *Venn diagrams*, to determine the validity of a given syllogism; it is the *structure* of the syllogism. Using Venn diagrams, we found that syllogisms of the structure *IAI-3* are valid and that those of the structure *EAE-4* are invalid. We pointed out earlier that there are 256 structures, in all, out of which only twenty-four are valid. All the others are known to be invalid. If we have the list of those that are valid, we need only evaluate a given syllogism to find its structure, then check whether it is one of those on the list. If it is, it is valid; if not, it is invalid. Given below is the complete list of those that are valid.[3]

AAA-1	*AAI*-1	*AAI*-3	*AAI*-4	*AI I*-1
AEE-2	*AOO*-2	*AEO*-2	*AI I*-3	*AEE*-4
AEO-4	*EAE*-1	*EIO*-1	*EAO*-1	*EAE*-2
EIO-2	*EAO*-2	*EAO*-3	*EIO*-3	*EAO*-4
EIO-4	*IAI*-3	*IAI*-4	*OAO*-3	

13.8 Ordinary Language and Arguments

We will start this section with a set of rules as checkpoints applicable to standard-form categorical syllogism. In a valid syllogism,

1. There should be only three terms (each used consistently).
2. The middle term should be distributed in at least one premise.
3. The terms in the conclusion should be distributed in the premises.
4. Both premises should not be negative.
5. If one of the premises is negative, the conclusion should be negative.
6. If both the premises are universal, the conclusion should be universal as well.

The above rules control the so-called standard syllogisms. Arguments as presented by researchers, even in the form of formal reports such as Ph.D. theses or journal publications, even those in scientific subjects (perhaps with the exception of mathematics and logic), seldom contain standard syllogisms. On the contrary, such reports are mostly formed with ordinary language. The reasons for this are understandable; ordinary language is more flexible, more resourceful, more interesting, and easier to use. But these attractions, themselves, may often lead the researcher away from the needed formality for correct thinking, resulting in botched arguments. The solution to such problems consists in translating the statements in ordinary language into the form of premises and conclusions and to arrange these, when possible, in the form of syllogisms, then to check their validity. Some of the problems in ordinary language and some of the possible solutions are briefly mentioned below in the form of examples.

Example 1: Educated persons do not live filthy; all doctors are educated, so we may expect no physician to be dirty.

The above passage is close to, but not quite, a syllogism. Some modifications are necessary to reduce it to a standard syllogism. It can be noticed that the words "doctors" and "physicians" are used synonymously; "live filthy" and "dirty" are also used synonymously; these inconsistencies in the use of terms should be rectified. With other minor changes, we may write the statement as

No educated persons are dirty persons	*E*	*M—P*
All doctors are educated persons	*A*	*S—M*
∴ No doctors are dirty persons	*E*	*S—P*

This is a standard *EAE-1* valid syllogism.

Example 2: All cats are meat eaters. A tiger, being a cat, eats meat.

Each of the premises in a standard syllogism should relate two classes and so should the conclusion. To reduce the above passage to standard form, "a cat" should be translated to the class of "cats"; and "eats meat" should be replaced from its present predicate form to a term designating the class of "meat eaters." With these necessary changes, the passage can be translated as

All cats are meat eaters	*A*	*M—P*
Tigers are cats	*A*	*S—M*
∴ Tigers are meat eaters	*A*	∴*S—P*

This is a structure *AAA-1* valid syllogism.

Example 3: Where there is smoke, there is fire. Finding no smoke here, we may not find any fire.

A legitimate syllogism can be formed even with singular things, instead of with classes. Such syllogisms are not of standard form. When an exactly similar term is made to repeat in all the three propositions of the syllogism, the reduction is known as *uniform translation*. The above passage may be reduced with such translation to

All places where there is fire are places where there is smoke: *A*

This place is not a place where there is smoke: *E*

∴ This place is not a place where there is fire: *E*

This is a syllogism with *AEE* mood with indefinite figure.

Example 4: My boy plays tennis. That is how he is athletic.

In this passage, the proposition that the boy is athletic is clearly the conclusion, but it is based on only one premise: that the boy plays tennis. Another premise, that "all those who play tennis are athletic," is missing. This is typical of an argument that is stated incompletely, the unstated part being assumed to be so evident that it is superfluous. Ordinary language is replete with such arguments. In logic, they are said to be *enthymematic*, and the argument is referred as an *enthymeme*. To check the validity of the argument, it is thus necessary to state the suppressed proposition and complete the syllogism. When the major premise is the one that is left unexpressed, the *enthymeme* is said to be one of the *first order*; when the minor premise is suppressed, it is of the *second order*; when the conclusion itself is suppressed, it is of the *third order*. To reduce the above statement to a syllogism, we need to supply the major premise and translate as

All those who play tennis are athletic	*A*	*M—P*
My boy plays tennis	*A*	*S—M*
∴ My boy is athletic	*A*	*S—P*

This makes it a structure *AAA-1* valid syllogism.

Example 5: Rama was either coronated as the future king or he was banished to the forest. He was not coronated as the future king; he was banished to the forest instead.

Disjunction is a condition of "either . . . or," not both. Between two alternatives, strictly as stated, if one is yes, the other is no. The above statement may be reduced to a standard-form syllogism as follows:

Either Rama was coronated as the future king, or Rama was banished to the forest.

Rama was not coronated as the future king.

∴ Rama was banished to the forest.

The above is referred to as a *disjunctive syllogism.*

Example 6: If there is rebirth, someone should have the experience of past birth. Since no one has such experiences, there cannot be rebirth.

This statement is of the form: "*if . . . then*"; thus, it is a hypothesis. The above statement can be translated as follows into a form known as a *hypothetical syllogism*:

If there is rebirth, then someone should have the experience of past birth.

No one has the experience of past birth.

∴ There is no rebirth.

Example 7: If children are responsible, parental discipline is unnecessary, whereas if children are irresponsible, parental discipline is useless. Children are either responsible or irresponsible. Therefore, parental discipline is either unnecessary or useless.

The above statement is aptly called a *dilemma.* Such arguments advanced by researchers may appear logical. That it is not so can be realized by comparing the dilemma with a *counterdilemma* as follows:

If children are responsible, parental discipline is well received and useful; whereas, if children are irresponsible, parental discipline is necessary to correct them. Children are either responsible or irresponsible. Therefore, parental discipline is either useful or necessary.

It should be noticed that in both the original dilemma and the counterdilemma, the fact that children are neither totally responsible nor totally irresponsible is suppressed. Dilemmas are known to sway an uncritical audience and, so, may serve well in political (and religious) propaganda. But the researcher, bound by logic, should learn to shun dilemmas.

13.9 References

1. S. J. Bachhuber, *Introduction to Logic* (Appleton-Century-Crofts, 1957), 8–9.

2. P. F. Strawson, *Introduction to Logical Theory* (Metheun and Co., University Paperback, 1971), 159.

3. Strawson, *Introduction*, 160.

13.10 Bibliography

Hurley, P. J. *A Concise Introduction to Logic.* 5th ed. Belmont, CA: Wadsworth Publishing Co., 1994.

Quine, M. V. *Methods of Logic.* 4th ed. Cambridge, MA: Harvard University Press, 1982.

Cohen, M. R., and E. Nagel. *An Introduction to Logic and Scientific Method.* London: Rutledge and Kegan Paul, 1961.

Winer, B. J. *Statistical Principles in Experimental Design.* New York: McGraw-Hill, 1971.

Beardsley, M. C. *Thinking Straight.* Upper Saddle River, NJ: Prentice-Hall, 1950.

14

Use of Symbolic Logic

If certain scientific elements—concepts, theories, assertion, derivations and the like—are to be analyzed logically, often the best procedure is to translate them into the symbolic language.

—*Rudolf Carnap*

A typical experimental researcher does not have to be proficient in symbolic logic, but some familiarity with it will be very helpful. When word combinations—phrases, clauses, sentences—can be interpreted in more than one way, it is possible for others to misunderstand the experimenter's statements and reports. In the analysis of complex experiments, premises that lead to conclusions, which in turn become premises that lead to the next level of conclusions, are not rare. After the advantage of using symbolic logic in such situations is pointed out, this chapter presents a brief review of *propositional logic*—that aspect of symbolic logic relevant to statements; the use of *truth tables* and *punctuations* are demonstrated. The chapter concludes with examples that connect ordinary language to symbolic language.

14.1 The Need for Symbolic Logic

That logic has an important part to play in experimental research is noted earlier. It is now time to distinguish between what is known as traditional logic and the topic we are going to outline below, namely, *symbolic logic* (also referred to as *mathematical logic*). The two logics differ in their means, their end being the same: coherent reasoning. In traditional logic, words, sentences, and punctuation are used as they are in any literature. Symbolic logic, on the other hand, uses symbols somewhat similar in appearance to those used in arithmetic. Traditional logic has usually been the intellectual province of philosophers, but symbolic logic is more a creation of mathematicians than of philosophers. It is no coincidence that two of the prominent logicians of the twentieth century, Bertrand Russell and Alfred North Whitehead, have been mathematicians turned philosophers.

Beginning about the middle of the nineteenth century, in the attempt to "reconstruct the foundations of mathematics," traditional logic was found inadequate. A more precise, accurate, and extensive language than that of traditional logic was found necessary, resulting in what came to be known as *symbolic logic*. In fact, symbolic logic, as it is now known, is a generic term. Under this broad heading, there are many variations, each an attempt to meet the needs of different disciplines. That branch of symbolic logic that is directly relevant to the experimental researchers is known as *propositional calculus* (or *propositional logic*). It deals with the analysis of unanalyzed but complete propositions and combinations of propositions, expressed not in words but, instead, in symbols.

> By way of indicating the merits of using symbolic language, Rudolf Carnap has this to say: "In this language, in contrast to ordinary word-language, we have signs that are unambiguous and formulations that are exact; in this language, therefore, the purity and correctness of a derivation can be tested with greater ease and accuracy. A derivation is counted as pure when it utilizes no other presuppositions than those specifically enumerated. A derivation in word-language often involves presuppositions that were not made explicitly, but which entered unnoticed. . . . A further advantage of using artificial symbols in place of words lies in the brevity and perspicuity of the symbolic formulas."[1]

While agreeing with Carnap that symbolic logic is a sharper tool, I do not intend to imply that any experimental researcher must become proficient in symbolic logic before doing experiments, or even before reporting his findings. What follows in this chapter is an outline introduction to symbolic logic (more precisely, propositional calculus) to indicate the ways in which it can be used to advantage by experimental researchers. This is not meant to be a learned exposition of the subject. The interested reader can access a large number of books at various levels. For those who realize the need, even marginally, the benefits of studying propositional calculus are well worth the time and effort.

14.2 Symbols in Place of Words

Firstly, we note that symbolic logic, however brief and crisp, is yet a language. The symbols used are, in a sense, the result of translating the ordinary language. The outcome of manipulating the symbols, according to preestablished rules, needs to be translated back into the ordinary language, before presentation to the public. As such, each ordinary language requires its own set of symbols to suit its syntax. Those that are presented here, obviously, have reference to the English language. Secondly, we note that what is referred to as a *proposition* is a statement, expressed in the form of a declarative, simple sentence. Thirdly, each statement used is either true or false, the truth-value of which is supposed to be known. Each statement may be symbolized by an alphabet of our choice. For example, the statement "Rosie is beautiful" may be symbolized by b (or any other letter in any alphabet). Out of such simple statements, compound statements are formed, in the process of developing the argument or forming the inference. The nature of connecting two (or more) statements, to develop a compound statement, that is, the *connective*, is also symbolized. Uses of some of the more common connectives, with symbolized statements, are summarized below. Depending on the function of the connectives, the combinations are known by different terms, the important ones of which are summarized in this chapter.

14.3 Conjunction

Two simple statements are said to form a *conjunction* when they have an "and" between them. The two simple statements are then known as *conjuncts*; for example, "*b* • *t*" is symbolic representation of "Rosie is beautiful" and "Rosie is tall," where *b* stands for "Rosie is beautiful," the dot, "•," stands for "and," and *t* stands for "Rosie is tall." Note, however, that any and every "and" does not necessarily function to form a conjunction. For example, "Rosie and Violet are sisters" is a simple (not a compound) statement; hence, "and" in this statement cannot be symbolized as "•," and the statement is not a conjunction.

On the other hand, the two statements "Rosie is beautiful" and "Rosie is tall" can be combined into "Rosie is beautiful and tall" without any change in meaning, and this change is economical. The symbolic representation "*b* • *t*," as before, is quite appropriate for this compound statement, which is a conjunction. The above case is one of repeating the subject part of the two conjuncts. If, instead, there are two simple statements in which the predicate part repeats for two different subjects, as, for example, in "Rosie is beautiful" and "Violet is beautiful," once again we may form a conjunction. "*R* • *V*" symbolizes the compound statement "Rosie and Violet are beautiful" in which *R* and *V* stand for "Rosie is beautiful" and "Violet is beautiful," respectively, and "•" stands for "and" without confusion.

By now, it may be obvious that the word "and" does not itself form the conjunction; it is the function it performs in the syntax that is relevant. If some other word performs a similar function, that word can be symbolized by "•" instead. For example, in the combined statement "Rosie is beautiful; also, Violet is beautiful," the word "also" performs the same function as "and." In the combined statement "Rose is beautiful; Violet is beautiful," even the semicolon is as good as "and." Likewise, in the English language, many other words and punctuation marks, depending on the syntax, can substitute for "and" (e.g., "yet," "but," "still," "moreover").

14.4 Truth Tables

Each simple statement being either true or false, that one which is not true is obviously false. In the case of the compound statement where two simple statements, say *p* and *q*, referred to as *truth-functional components*, form a conjunction, they form four possible combinations:

1. p is true; q is true.
2. p is true; q is false.
3. p is false; q is true.
4. p is false; q is false.

The conjunction itself, in its entirety, as expressed, to be true, should have both p and q to be true. For instance, for the statement "Rosie is beautiful and tall" to be true, Rosie should be both beautiful and tall. If Rosie is (a) beautiful but not tall, (b) not beautiful but tall, or (c) neither beautiful nor tall, the statement of the conjunction is false. So, among the four combinations listed above, only (1) is true; all others are false.

Expressing such facts in tabular form, as *truth tables*, is fairly common in propositional logic. Table 14.1 shows the truth table for the case of conjunction that we discussed above, using *T* for true and *F* for false.

Table 14.1 *Truth table for conjunction.*

p	q	$p \cdot q$
T	T	T
T	F	F
F	T	F
F	F	F

As can be observed, Table 14.1 inevitably follows from the way the dot symbol "•" is meant to operate. Hence, conversely, it may be taken as the definition of "•" in conjunctions.

14.5 Disjunction

Two simple truth-functional statements are said to form a *disjunction* when they are connected by an "or" or another word that performs the same function in the syntax. The two simple statements are then known as *disjuncts.* Each such statement is represented by a symbol of our choice, and the connective is symbolized by "v," referred to as a *vedge* or *vee*. Thus $p \vee q$, as a compound statement, is a disjunction. The word "or" in the English language has many inflections, capable of different interpretations, like (1) "either p or q, but not both" (e.g., "My boy is wearing either a blue cap or a red cap), or (2) "either p or q, possibly both" (e.g., "To qualify for the field trip, the student should get an "A" in English or Math"). The latter interpretation may further be understood as (3) "at least one of p and q," and, in this form, it subsumes the interpretation (1) as well. This, indeed, is the scope of "v," and Table 14.2 shows the truth table for the disjunction.

Table 14.2 *Truth table for disjunction.*

p	q	$p \vee q$
T	T	T
T	F	T
F	T	T
F	F	F

When either the subject or predicate part of the disjunct is the same, economy of the language shows in their fusion: for example, "The newborn is a boy" (B) and "The newborn is a girl" (G), are usually combined as "The newborn is either a boy or a girl" ($B \vee G$); and "The student is studying" (s) and "The student is playing" (p) are combined as "The student is studying or playing" ($s \vee p$).

The literal presence of "or," it should be noted, is not necessary for a disjunction; it is the sense that counts. For instance, the statement "Unless you are busy" (b), "we play tennis" (t) can be symbolized as $b \vee t$, even without an "or." The sense of the statement can be understood as "Either we play tennis or you are busy."

Having said this, we should note that the "if . . . then" condition may be used in several different senses. In the above example, if Vishwa now is 5′8″, then with an additional four inches, his becoming 6′ is factual; it is logical. But all "if . . . then" statements need not be logical. For instance, the statement "If I were a rich man, then I could be a philanthropist" is not logical; it is only a wish. But the statement is conditional, nonetheless. Another subtle observation is required before we prepare the truth table for the conditional statement by way of defining "⊃." In addition to the previous two, we may add some more conditional statements:

If he is a widower (*p*), then his wife died (*q*).

This is the definition of a widower.

If he hits harder (*p*), the ball will go over the fence (*q*).

This is a causal (or empirical) relation.

If you lie, you will go to hell.

This is an admonition, an imperative.

The five statements listed in this section are all conditional; many more can be added to the list. Now we may ask the question, What is the common factor that makes all these statements conditional, though the degree of logic in these statements varies? To answer this question, we should understand the scopes and limitations of the symbol "⊃." When used as $p \supset q$, it implies that if *p* is true, *q* is true. It does not say anything about either *p* or *q* being factual.

Now, suppose we transform the conditional statement "if *p* then *q*" to the format of a conjunction. Furthermore, making use of the negation symbol, $-q$ stands for "*q* is false." Now, *p* being true, and *q* being false, which can be symbolized as $p \bullet \sim q$, makes this conditional statement false. In another step, if we take $p \bullet \sim q$ to be false, its negation, namely, $\sim (p \bullet \sim p)$, can be taken to be true. And this last expression, $\sim(p \bullet \sim q)$, symbolizes the truth or falsehood of the combination for the corresponding truth or falsehood of *p* and *q*.

14.6 Negation

Negation is denial, contradiction
such words or phrases as “no,” “n
and “it is not the case that.” Th
truth-functional statement is “~,’
however, may have different variat
tence “All Americans are good pec
ous statements, such as “Not all
“Some Americans are not good pe
Americans are good people,” and
people,” are statements for which ~
that, unlike in conjunction and dis
functional statements are involved,
one element, the truth table of whic
(see Table 14.3).

Table 14.3 *Truth table for negation.*

p	$\sim p$
T	F
F	T

14.7 Conditional Statements

Two truth-functional statements linke
tion are said to form together a *cond*
hypothetical or *implicative*) statement.
that fills the gap between the “if” an
antecedent (or *implicans*), and that wl
known as the *consequent* (or *implicate*).
more inches, then he will be six feet
more inches” is the antecedent, and “he
consequent. The antecedent and the co
necessarily) symbolized by p and q, re
bined conditional statement is then sy
symbol “⊃” is known as a *horseshoe*.

14.6 Negation

Negation is denial, contradiction, or falsification expressed by such words or phrases as "no," "not," "not so," "it is false that," and "it is not the case that." The symbol used for negating a truth-functional statement is "~," known as a *kurl.* Negation, however, may have different variations. For instance, if the sentence "All Americans are good people" is symbolized as *A*, various statements, such as "Not all Americans are good people," "Some Americans are not good people," "It is not true that all Americans are good people," and "All Americans are not good people," are statements for which ~*A* is equally applicable. Note that, unlike in conjunction and disjunction, wherein two truth-functional statements are involved, we have, in negation, only one element, the truth table of which, serves as definition of "~" (see Table 14.3).

Table 14.3 *Truth table for negation.*

p	$\sim p$
T	F
F	T

14.7 Conditional Statements

Two truth-functional statements linked by an "if . . . then" relation are said to form together a *conditional* (often also called a *hypothetical* or *implicative*) statement. The elemental statement that fills the gap between the "if" and "then" is known as the *antecedent* (or *implicans),* and that which follows the "then" is known as the *consequent* (or *implicate*). In "If Vishwa grows four more inches, then he will be six feet tall," "Vishwa grows four more inches" is the antecedent, and "he will be six feet tall" is the consequent. The antecedent and the consequent are usually (not necessarily) symbolized by p and q, respectively, and the combined conditional statement is then symbolized as $p \supset q$; the symbol "$\supset$" is known as a *horseshoe.*

Having said this, we should note that the "if . . . then" condition may be used in several different senses. In the above example, if Vishwa now is 5′8″, then with an additional four inches, his becoming 6′ is factual; it is logical. But all "if . . . then" statements need not be logical. For instance, the statement "If I were a rich man, then I could be a philanthropist" is not logical; it is only a wish. But the statement is conditional, nonetheless. Another subtle observation is required before we prepare the truth table for the conditional statement by way of defining "⊃." In addition to the previous two, we may add some more conditional statements:

If he is a widower (*p*), then his wife died (*q*).

This is the definition of a widower.

If he hits harder (*p*), the ball will go over the fence (*q*).

This is a causal (or empirical) relation.

If you lie, you will go to hell.

This is an admonition, an imperative.

The five statements listed in this section are all conditional; many more can be added to the list. Now we may ask the question, What is the common factor that makes all these statements conditional, though the degree of logic in these statements varies? To answer this question, we should understand the scopes and limitations of the symbol "⊃." When used as $p \supset q$, it implies that if *p* is true, *q* is true. It does not say anything about either *p* or *q* being factual.

Now, suppose we transform the conditional statement "if *p* then *q*" to the format of a conjunction. Furthermore, making use of the negation symbol, $-q$ stands for "*q* is false." Now, *p* being true, and *q* being false, which can be symbolized as $p \bullet \sim q$, makes this conditional statement false. In another step, if we take $p \bullet \sim q$ to be false, its negation, namely, $\sim (p \bullet \sim p)$, can be taken to be true. And this last expression, $\sim(p \bullet \sim q)$, symbolizes the truth or falsehood of the combination for the corresponding truth or falsehood of *p* and *q*.

We shall use these steps as keys to prepare the truth table for the conditional statements as in Table 14.4.

Table 14.4 *Preparing truth table for conditional statements.*

p	q	$\sim q$	$(p \bullet \sim q)$	$\sim (p \bullet \sim q)$
T	T	F	F	T
T	F	T	T	F
F	T	F	F	T
F	F	T	F	T

Now, omitting the intermediate steps used as keys, and noting that the last column asserts the truth-value of the statements $p \supset q$, we may write the truth table for the conditional statement; this, in turn, serves as the definition of "$\supset$" (see Table 14.5).

Table 14.5 *Truth table for conditional statements.*

p	q	$p \supset q$	
T	*T*	*T*	(1)
T	*F*	*F*	(2)
F	*T*	*T*	(3)
F	*F*	*T*	(4)

Visiting the Table 14.5 afresh, we may make the following observations relative to the four cases:

(1) If p is true, q is true: conforms to $p \supset q$; hence, true

(4) If p is not true, q is not true: conforms to $p \supset q$; hence, true

(2) If p is true, but q is not true: does not conform to $p \supset q$; hence, false

Case (3) is not so obvious as the others, but having followed the logic of its derivation with mathematical precision, we need to take it in good faith. As a parallel, not meant to prove the case (3), but only in way of supporting it, we may consider the statements of a tennis player:

Before the match: "If I play very well, I can win this match."

After the match: "I did not play very well, but I won the match anyway."

Having defined the conditional symbol "$\supset$," we need to note a few varieties, even derivations, some of which may appear to contradict the definition. Firstly, the literal presence of "if" and "then" is not a requirement; what is required is that the sense of such a condition be conveyed, despite the absence of these words, or by use of other words instead. The sense of "If I were a rich man, then I would be a philanthropist" can be conveyed by other statements, such as

"Had I been a rich man, I could be philanthropic."

"In case I were a rich man, I would be a philanthropist."

"If God made me rich, I should follow philanthropy."

Secondly, the place of antecedent and consequent may get interchanged without altering the meaning; for instance, "I would philanthropize, if God made me rich." Thirdly, the antecedent to consequent relation may not always be one to one. For a given consequent, there may be more than one condition required, meaning several antecedents together lead to a particular consequent. The statement "A good soccer player needs foot speed, skill, and stamina" is meant to convey that any one of these qualities is necessary but not sufficient. A conditional statement for one such quality is "If one has foot speed, then one can become a good soccer player." A slightly different situation is that in which there are several antecedents, each sufficient, leading to the same consequent. Jogging, bicycling, or playing tennis may each be sufficient to meet the needs of a person who is fitness conscious. A conditional statement relative to one such activity would be "If one plays tennis, then one can be physically fit."

In summary, we can say that the conditional relation $p \supset q$ subsumes several variations. It can be adapted to many situations. In some situations, perhaps, considerable moderation in relating the two elemental statements may become necessary.

14.8 Material Implication

Here is a situation in which the scope and significance of "⊃" need to be reassessed. We have noted that truth content of conditional statements can be different in nature: logical, causal, definitional, and so forth. The common factor, though, is that there is a relation between the contents of the antecedent and the consequent, in the sense that the consequent is relevant to the antecedent. But nothing prevents making a statement using "if . . . then" in the English language, in which the consequent is totally irrelevant to the antecedent; for instance, "If it rains tomorrow, then an elephant is a big animal." Such a statement, in which the relevance—an obvious connection between the antecedent and the consequent—is missing, is known as *material implication*.

In the process of translating any "if . . . then" statement to a symbolic language, using "⊃," unfortunately, we need to treat any conditional statement as a material implication, as if one is functioning like a robot. Much as we like to ascribe to propositional logic the simplicity and precision of elementary arithmetic with its operational symbols "+," "−," "×," and "÷," we need to be aware that the conditional statement in symbolic format need not necessarily have a meaning, let alone a truth function. Relevance between the antecedent and the consequent is an essential requirement to make the conditional statement meaningful, and to decide this criterion, we still remain dependent on ordinary language. This may be a serious limitation of the scope of "⊃."

14.9 Punctuation in Symbolic Logic

There is no sentence without *punctuation*—at least a period ("."), exclamation mark ("!"), or question mark ("?")—at the end of it. When there is more than one phrase or clause within a sentence, additional punctuation becomes necessary to make the meaning clear. Compare the sentence "Anthony tried to force a crown upon Caesar which Caesar after a little coyness and in face of the manifested displeasure of the crowd refused" with its punctuated version: "Anthony tried to force a crown upon Caesar, which Caesar, after a little coyness and in face of the manifested displeasure of the crowd, refused."

The jumble, leading to many guesses, is made clear by inserting commas, as required. If this is the case with a statement in words, then with a statement in numbers, it is worse. Compare

$6 \times 3 + 8 \div 4 - 2$

with

$(6 \times 3) + (8 \div 4) - 2$, which equals 18,

or with

$[(6 \times 3) + 8] \div (4 - 2)$, which equals 13.

The use of parentheses and brackets here, like the use of commas previously, is necessary to clarify the intention of the statement, which is the function served by punctuation. As with numbers, propositional logic mostly uses parentheses "( )," then occasionally brackets "[ ]," and rarely braces, "{ }," in that higher order. When used among truth-functional statements, which we will do later, the meanings of punctuation are exactly similar to those used in elementary arithmetic, as shown above.

14.10 Equivalence: "Material" and "Logical"

Equivalence in propositional logic, perhaps more than any other notion, has the closest analogy to that in mathematics, and the symbol used, "≡," is the same. In terms of truth-value, however, two variations of equivalence are recognized in propositional logic: (1) material equivalence, and (2) logical equivalence.

14.10.1 Material Equivalence

When two statements are both true and both false, those two statements are said to be in *material equivalence*, symbolized by a "≡" between the two statements. This is shown in truth table 14.6, which also serves as the definition of "≡".

Table 14.6 *Truth table for material equivalence.*

p	q	$p \equiv q$
T	T	T
F	F	T
T	F	F
F	T	F

Further emphasis of $p \equiv q$ may be mentioned as "p if and only if q" or "q if and only if p," both of these being interchangeable; because of this, the material equivalence is said to be *biconditional* or *bifunctional*. What distinguishes material equivalence is the situation in which the two statements should both be true or false; they need not be speaking about the same fact. For instance, in $p \equiv q$, p may be the symbol for "Peacock is beautiful" and q the symbol for "Elephant is a big animal."

14.10.2 Logical Equivalence

Two statements are said to be *logically equivalent* when those statements express the same truth-value, or the same meaning, or the same fact; the sameness is what is required. Such statements are known in logic as *tautologies*. Thus, if p stands for "Rosie is beautiful," then $\sim p$ stands for "Rosie is not beautiful," and $\sim(\sim p)$ stands for "It is not true that Rosie is not beautiful," making p logically equivalent to $\sim(\sim p)$. The truth table for logical equivalence is shown in Table 14.7.

Table 14.7 *Truth table for logical equivalence.*

p	$\sim p$	$\sim(\sim p)$	$p \equiv \sim(\sim p)$
T	F	T	T
F	T	F	T

It is interesting to note that ~(~*p*) conforms to the popular saying "Double negative makes an affirmative." Even more so, the sign "~" here performs the same function as a minus sign ("–"), in mathematics, wherein "minus times minus," "– (–)," results in a plus, "+."

14.11 Application of Symbolic Logic

Now, with all the preparation we have had throughout this chapter, we may be ready to experience the application of symbolic logic. The question to ask is, What do we apply it for? meaning, What purpose is symbolic logic expected to serve? According to Carnap, quoted earlier, the first step is to be able to translate statements expressed in ordinary English language into their symbolic form, also known as *formal language*. In order to do that, we require the symbols ".," "v," "⊃," and "≡." Because of their operational functions, these are known as *operators*. Symbolizing each simple, elemental statement, and using the operators as needed, accompanied by the use of punctuation marks, when necessary, we should be able to build more complex (complicated, if necessary) statements. Once this stage is reached, the operators come into play. Simplification of the complex statement, done in steps, following the hierarchy of parenthesis, brackets, and braces, if there are any, should lead us to the truth-value of the entire statement. The truth that emerges from such simplification may be an answer to a question posed by the complex statement or an inference obtained from various observations, each observation expressed as an individual statement or a proof of validity, or otherwise, of a hypothesis.

14.11.1 Ordinary Language to Symbolic Language

Example 1: The sentence "If a small country is bombed by the United Kingdom, and trade sanctions are imposed by the United States, it is not in the power of the United Nations to protect the small country" may be symbolized as

$$\sim[(K \,.\, S) \supset N]$$

where:

K = A small country is bombed by the United Kingdom

S = Trade sanctions are imposed (on the small country) by the United States

N = The United Nations has the power to protect the small country

Example 2: The sentence "I will travel by bus if and only if the distance is less than two hundred fifty miles or the airfare is more than $200" may be symbolized as

$$B \equiv (D \vee F)$$

where:

B = I will travel by bus

D = The distance is less than two hundred fifty miles

F = The airfare is more than $200

14.12 Validity of Arguments

As we have hinted occasionally, though we have not explicitly said so, there are some similarities between the concerns of a trial attorney and those of a research scientist: both need to present arguments, each as a way of proving his case. The difference is in the circumstances. The trial attorney presents his questions and cross-examinations, addressed to the witnesses, for the most part vocally in front of the judge and the jury. The experimental scientist presents his argument in the form of a discussion of experimental results, mostly in written form. The judge and the jury for him are those who happen to read his paper (or article, dissertation, or thesis) because of some common interest, even though such interest may be evinced after a considerable lapse of time. Occasionally, such papers will be presented before a live audience, usually not a very critical one, in seminars or scientific conferences or colloquiums. Another difference between a researcher and a trial attorney is that the trial attorney builds his arguments on the circumstantial evidence obtained by way of the answers and confessions of the witnesses and the material evidence of exhibits. The experimental researcher, on the other hand, builds his arguments based on the outcome of observed or recorded experimental results, the cause-and-effect relations between or among the parametric variables involved.

What is common to both the trial attorney and the experimental researcher is that in the process of building the argument, several pieces of evidence may need to be put together with logical connections to form the conclusion. It is not uncommon in experimental investigation, as much as in judicial trial, that arguments are reached, not in one leap, but in several steps, with inference from the first step becoming the premise for the second step, and inference from the second step becoming the premise for the third step, and so on, until the final conclusion is reached. For an inference at any particular step to be valid, it should be based on true premises. For the final conclusion to be valid, the inferences at each step should be valid. And the test for the validity of arguments is facilitated enormously if the statements are symbolized. That is the use of symbolic logic for the experimental researcher.

14.13 Reference

1. R. Carnap, *Introduction to Symbolic Logic and its Applications* (New York, NY: Dover Publications, 1958), 2.

14.14 Bibliography

Barry, V. E. *Practical Logic*. New York: Holt, Rinehart and Winston, 1976.

Carnap, R. *Introduction to Symbolic Logic and Its Applications*. Mineola, NY: Dover Publications, 1958.

Copy, I. M., and C. Cohen. *Introduction to Logic*. 8th ed. Indianapolis, IN: Macmillan Publishing Co., 1990.

Kelly, D. *The Art of Reasoning*, Expanded ed. (with symbolic logic). New York: W. W. Norton and Co., 1990.

Russell, B. *Principles of Mathematics*, 2nd ed. W. W. Norton and Co., 1938.

Strawson, P. F. *Introduction to Logical Theory*. London: Mathuen and Co., 1952.

Part V: Probability and Statistics for Experimental Research

15

Introduction to Probability and Statistics

To understand God's thoughts we must study statistics, for these are the measures of his purpose.

—*Florence Nightingale*

This chapter provides the reader a brief but comprehensive study of the basics of probability and statistics. It starts by pointing out that the kind of knowledge derived from experiments is necessarily probabilistic in nature. After the words "probability" and "statistics" are defined separately, the nature of their mutual relation is indicated. The need and reasons for assigning numerical values of probability to several a priori situations based on relative frequency are shown. Methods for calculating, with variations, the statistical values for the *central tendency* and *dispersion* of numerical data are illustrated. Then, the different ways of presenting statistical data—tabular and graphical—are detailed. The final part of the chapter points to the importance of the normal distribution curve, ending with some typical distributions that are not normal.

15.1 Relevance of Probability and Statistics in Experimental Research

A misgiving that could possibly cross the mind of an experimenter is, Why, in a book on experimental research, do we need to study probability and statistics? The basic answer is, Because what we do in experimental research is observe something, an effect, happening under a given set of conditions. And, because under the same set of conditions, created at different times, maybe elsewhere, it is not certain that we would see exactly the same effect as no proof can be advanced, independently of more experiments. Bertrand Russell forcefully projects that most of our knowledge obtained by observation, which is what we do in experimental research, is probable knowledge, not certainty. He cites such commonplace observation as the sun rising tomorrow:

> If we are asked why we believe that the sun will rise tomorrow, we shall naturally answer, "Because it has always risen everyday. . . ." If we are challenged as to why we believe that it will continue to rise as heretofore, we may appeal to the laws of motion. . . . The only reason for believing that the laws of motion will remain in operation is that they have operated hitherto. . . . [T]he sunrise is merely a particular case of fulfillment of the laws of motion. . . . [A]ll such expectations are only *probable*; thus, we have not to seek for a proof that they *must* be fulfilled, but only for some reason in favor of the view that they are likely to be fulfilled.[1]

Being thus confined to observation in experimental research, hence, to the world of probability, we may still have the freedom to seek an answer to the question, How far removed is the object of our observation, namely, the effect, when it is quantitative, from what it could be if it were certain, even though certainty remains an unreachable ideal? Dependable answers to this question, suitable for various situations, are available in the form of probability theory, which in turn, forms the base for statistical methods. And, wherever a large bulk of numbers is involved, as in experimental methods, the use of statistical methods is unavoidable. S. Goldberg writes, "The theory of probability, as the foundation upon which the methods of statistics are based, should command the attention of those who want to understand as well as apply statistical techniques. Probabilistic theories, making explicit reference to the nature and effects of chance phenomena, are the rule rather than the exception in the physical and biological sciences."[2]

Let us now consider a practical application. Attempts to obtain by experimental means the relation between two variables are all too common in science and technology. In drawing a graphical relation between an independent variable x and a dependent variable y, seldom do we have all the points on a smooth curve. We draw a line that is approximately close to most points. When the points show a considerable amount of deviation from a possible smooth curve, we need to find the "best fit" for the points. The method used for doing this, and to obviate the guesswork, involves probability and statistics. Once we have such a relation between x and y, we may predict the value of y for any given new value of x (and vice versa) within the range. Suppose after getting such a value of y, a question is asked: Is this relation between the new values of x and y exact? No person with scientific training would say, Absolutely. Instead, we may expect an answer somewhat like, We may expect it to be reasonably good. The words in this statement "expect" and "reasonably" have their bearing in probability and statistics.

Many such concrete situations can be presented to justify the need to study probability and statistics as part of the preparation for experimental research. Thus, the study of probability and statistics is not an addendum to scientific research; besides being an essential tool, these fields are inextricably mixed in with the stuff of science. Many books are available for the reader to choose from, dealing at various levels, with all aspects of probability and statistics—philosophical, logical, and mathematical (quantitative). No attempt is made here to deal with these aspects in anything more than a bare outline. The philosophy of probability in this book is restricted to a discussion of the work of Russell, a well-known, British, twentieth-century philosopher. The logical aspect, though in a sense the most significant, is confined to illustration with a simple example. The quantitative aspect—better known and more often used than other aspects of probability—is synoptically dealt with but limited to basics. We turn first to the definitions of, and mutual relation between, probability and statistics.

15.2 Defining the Terms: Probability and Statistics

15.2.1 Probability

The word "probability" is too familiar to need examples; we use it quite freely, often knowing intuitively what we mean. Even when we attach a quantitative qualification, such as "a 70 percent chance of rain today," we feel fairly comfortable. Probability variously refers to, among other things,

- The degree of rational belief
- The quality of being likely (to happen)
- The field of partial knowledge and partial ignorance
- The study of chances

Pressed to define the word "probability," we face a hopeless task. Nonetheless, quantitative analysis of events on the basis of probability abounds, some of which follows later.

15.2.2 Statistics

The word "statistics" seems to have found usage among scientists, after that of probability, and prevails in such expressions as, "It is statistically true . . ." and "Statistics show . . .," and it manifests itself in a well-known branch of physics called statistical mechanics. The word is variously understood to refer to, among other things,

- A study of quantitative facts
- A method of interpreting measurements
- Comprehensive knowledge gained from a mass of isolated facts
- A method of gaining inference from a large quantity of data

> M. Richardson gives us a broader explanation: "It is difficult to give a precise definition of statistics. Speaking loosely, statistics is concerned with coming to general conclusion, whose truth is more or less probable, concerning large classes of phenomena, by means of the study of large numbers of observations. It is therefore of the greatest importance in connection with the inductive logic of experimental methods."[3]

The word "statistics" is said to have originated from the word "state," implying a large number (of people). What characterizes statistics is, thus, large numbers; it may be of quantities, measurements, opinions, votes, guesses, predictions, and other such entities.

15.3 Relation between Probability and Statistics

In this section we intend to indicate the nature of the relation between probability and statistics. Suppose we want to measure the computer literacy of sixth graders in the United States. We cannot literally test every sixth grader in the country. We can, and normally do, select a small number of *representative samples* of sixth graders and test them for the prescribed skills. The information gathered from such tests, known as *data*, is analyzed by statistical methods. No matter how precise the calculations involved in the analysis, the inference from the test cannot be 100 percent accurate because only a relatively small number of sixth graders was tested. Any statement that we make based on the outcome of the statistical analysis of the information we gather from the samples, thus, will necessarily have an element of uncertainty. In other words, the statements will not have a certainty- (or truth-) value; instead, they can only have *probability-value*. The statements that have only probability-values are often referred to as *statistical facts* or *statistical truths.*

Having said this, we need to note that the relation between probability and statistics prevails not necessarily with such large numbers as all sixth graders of the United States. They apply to even relatively small numbers, such as the sixth graders in a state or city, or even in a particular school. Suppose a *statistical fact* states that the sixth graders of Winslow School are more computer literate than those of Rodman School, both in New Bedford. Further, suppose a sixth grader is randomly selected from Winslow School, and another sixth grader is randomly selected from Rodman School. Statistics suggests that the particular student from Winslow is likely to be better than the particular one from Rodman, but this is only a probability; there is no surety that it will be so. One should not be surprised if the student from Rodman is found to be better than the other from Winslow. This possibility cannot be ruled out simply on the grounds that statistics suggests otherwise. Besides noting that the statement of statistical truth is thus a probability and not a certainty, we need to note the intimate, in fact, nearly inseparable, relation between probability and statistics.

15.4 Philosophy of Probability

Discussing the relation between truth and our knowledge, Russell alerts us that all of our knowledge of truth is infected with *some* degree of doubt:

> [B]oth as regards intuitive knowledge and as regards derivative knowledge, if we assume that intuitive knowledge is trustworthy in proportion to the degree of its self-evidence, there will be a gradation in trustworthiness, from the existence of noteworthy sense-data and the simpler truths of logic and arithmetic, which may be taken as quite certain, down to judgments which seem only just more probable than their opposites. What we firmly believe, if it is true, is called *knowledge*, provided, it is intuitive or inferred (logically or psychologically) from intuitive knowledge from which it follows logically. What we firmly believe, if it is not true, is called *error*. What we firmly believe, if it is neither knowledge nor error, and also what we believe hesitatingly, because it is, or is derived from, something which has not the highest degree of self-evidence, may be called *probable opinion*. Thus, the greater part of what would

commonly pass as knowledge is more or less probable opinion.[4]

15.5 Logic of Probability and Statistics

One should not lose sight of the fact that the concern expressed through such philosophy has direct relevance to experimental research in that any research has, as its goal, the enhancement of knowledge. Though knowledge for its own sake is desirable, knowledge, to run many practical affairs of life, is essential. Obtaining good crops from agriculture, curing certain diseases with medicine, administering criminal justice, getting a better quality of manufactured items, running an insurance business, and forecasting tomorrow's weather are but some of the more visible among the essential aspects of life, and in each one of these, the widespread use of probability and statistics is obvious. Considering weather forecasting—for instance, how wind current, humidity, temperature, weather in the neighborhood, local geography, and other factors interact to result in cloud and drizzle or bright and sunny conditions—predictions cannot be obtained through strictly derived and proven equations. The factors are uncontrollable; they are too many, and their interactions are too complex to subject them to an experimental investigation. The meteorologist's expertise, instead, consists of knowing, under the prevailing conditions approximated, how often in the past a day of cloud and drizzle followed and how often it did not. He resorts to probability and statistics. The logic here consists of having on hand a large number of factors, the interactions among which are too complex, yet to be able to develop a reasonable degree of belief in the event yet to happen. The degree of belief is assigned a number; the meteorologist, for instance, saying that there's a 70 percent chance of rain.

15.6 Quantitative Probability

The use of probability goes as far back in human history as the beginning of reasoning, for reasoning, in the broadest sense, is the search for the "probable" among the "possible." But not until the middle of the seventeenth century was the significance of probability made explicit by two mathematicians, Blaise Pascal

(1623–1662) and Pierre de Fermat (1608–1665), who by quantifying probability, that is, assigning a number to it in various circumstances, gave it a life of its own. The potentialities of probability, long dormant, were released in the form of quantities to develop into a new branch of mathematics, soon to find applications in many aspects of life, in turn changing the process of thought itself. This phase, namely, the *quantitative*, that is most often, and unless otherwise qualified, taken to be the purpose and the body of probability.

The subject is highly developed, suitable for application to several conceivable complexities of practical life. There are a host of books for the beginner to delve into as deeply as he or she likes or, if need be, to find parallels to his or her applications on hand and the required analysis. All that can be done here is to introduce the experimental researcher to the kind of analysis involved in assigning certain numbers to a few simple and basic situations; otherwise, wonder and guesswork could hold sway.

Broadly, there are two views, not wholly different from each other, often referred to as *theories of probability*: a priori theory and relative frequency theory.

A brief discussion of each follows:

15.6.1 A Priori Theory

Also known as *theoretical probability*, *a priori theory* consists of assigning a number, often called a *probability coefficient*, to hypothetical events that may or may not occur, but for which the various possibilities of the occurrence are known. No trials or experiments are required in obtaining the numbers here, this number being based only on rational thought of imagined circumstances. A number so obtained may later be checked for accuracy—closeness to expectation—by actualizing the circumstances of the events. A few examples follow:

1. Single independent event:

   Example: tossing a coin

   If one were to toss an unbiased coin, say, one hundred times, the coin is expected to land heads up fifty times; the other fifty times it is likely to land tails up. This is so because, besides these two, there are no other possibilities. Each toss in the above series of one hundred is independent of any other, meaning its outcome is not influenced by the previous or subsequent tosses. Therefore, we (rationally) assign the probability of heads landing up to be 50 ÷ 100 or 0.5.

2. Joint independent events:

   Example: rolling (two) dice

   If two dice are rolled together, the way one of those lands has no influence whatsoever on the way the other does; the two occurrences are, thus, independent. Suppose we want to know the probability of rolling a twelve (with two dice). This can happen only if the first die lands six up and the second one also lands six up. The first die has six faces, hence, six possibilities on landing; one of these is the number six. So, the probability of the first die landing six up is 1/6. The same logic applies to the second die, and the probability of six landing up is again 1/6. Then, the combined probability of the two independent occurrences is 1/6 × 1/6 = 1/36. This is known as the *product theorem of probability.* It holds not only for two, but for any number, of independent occurrences. Symbolizing, we may write,

$$P_{(a \text{ and } b \text{ and } c\ldots)} = P_{(a)} \times P_{(b)} \times P_{(c)} \ldots$$

   where P stands for "probability of," and a, b, c . . . stand for corresponding occurrences.

3. Dependent (subsequent) events:

   Example: drawing colored balls from a bag

   We imagine a bag containing one hundred otherwise identical balls of different colors as below:

Black	10
White	20

Red	30
Yellow	40

If a single ball is drawn at random, the probability of that ball being any of the colors is as below:

Black	10/100 = 1/10
White	20/100 = 1/5
Red	30/100 = 3/10
Yellow	40/100 = 2/5

Suppose the first ball drawn is black. If it is put back into the bag, there are one hundred balls once again, and the probabilities for the next draw are the same as above. The events are independent. If, instead, the first ball drawn is not put back into the bag, the probabilities of all subsequent draws will be altered. They become dependent because now there are only nine black balls among a total of only ninety-nine. The probabilities for drawing a second ball of any of the colors are

Black	9/99
White	20/99
Red	30/99
Yellow	40/99

If this second ball is also black and is not put back into the bag, and a third ball is drawn, the new probabilities are

Black	8/98
White	20/98
Red	30/98
Yellow	40/98

Now, if we want to know the probability of drawing black balls, three consecutive times, starting with one hundred balls as before, it is:

$$10/100 \times 9/99 \times 8/98 \quad \text{or} \quad \frac{(10 \times 9 \times 8)}{(100 \times 99 \times 98)}$$

Symbolizing, we may write, using "if" for dependence,

$$P_{(a \text{ and } b \text{ and } c...)} = P_a \times P_{(b \text{ if } a)} \times P_{(c \text{ if } a \text{ and } b)} \times ..$$

4. Mutually exclusive but alternate occurrences:

Example: tossing two coins for either two heads or two tails

If two coins are tossed together, the events of getting two heads and two tails are mutually exclusive. If the probability of getting *either* two heads *or* two tails, *no matter which*, is required, then the probabilities of both need to be added. The probability of the first coin coming up heads is 0.5 and that of the second also coming up heads is 0.5. Hence, the probability of two heads together is 0.5 × 0.5 = 0.25. Similarly, the probability of two tails together is also 0.25. So, the (combined) probability of getting either two heads or two tails is 0.25 + 0.25 = 0.5.

Symbolizing, we may write,

$$P_{(a \text{ or } b)} = P_a + P_b$$

which is known as the *Additive Theorem.* This can be further confirmed as below:

Various possibilities of tossing the two coins, marked 1 and 2, and symbolizing H for heads and T for tails are as follows:

$$\underline{H_1}—\underline{H_2},\ H_1—T_2,\ T_1—H_2,\ \underline{T_1}—\underline{T_2}$$

5. Various alternative joint occurrences:

   Example: rolling (two) dice for seven

   In the previous instances of rolling two dice for twelve, only one of the thirty-six possibilities of rolling the two dice together met the requirement; hence, the probability for that was 1/36. If the number required is seven instead, many occurrences meet the requirement; they are one and six, six and one, two and five, five and two, three and four, and four and three.

   Tabulating all the possibilities of rolling, we can identify these occurrences (underscored), as shown in Table 15.1.

Table 15.1 *Rolling two dice for seven.*

1—1	2—1	3—1	4—1	5—1	6—1
1—2	2—2	3—2	4—2	5—2	6—2
1—3	2—3	3—3	4–3	5—3	6—3
1—4	2—4	3—4	4—4	5—4	6—4
1—5	2—5	3—5	4—5	5—5	6—5
1—6	2—6	3—6	4—6	5—6	6—6

Since six alternatives fulfill the requirement, among the thirty-six rolls of the two dice jointly, the probability is 6/36, or 1/6.

15.6.2 Relative Frequency Theory

In contrast to a priori theory, the *relative frequency theory* depends, not on reasoning, but on observation in the form of data. There are two variations to this theory. If the data was derived in the form of the effect or outcome of experiments, it is referred to as the *experimental method.* If, instead, the data was the result of a survey or study of a statistical nature, it is referred to as the *inferential method.* The experimental method is of special significance to us, and we shall study its application in a later chapter on inference. For demonstrating the inferential statistical method, on the other hand, we may imagine an educator who is interested in knowing the performance of the tenth-grade students within a school district on a common "standard" test in, say, mathematics; his purpose is to know how frequently students score 95 percent or above. If he obtained data about two hundred students selected *at random*, and if forty-six of those scored 95 percent or above, then the probability of this occurrence would be 46/200 or 0.23. Incidentally, in such studies, the two hundred students are referred to as the *class*, and "scoring 95 percent or above" is referred to as the *attribute*.

It shall be noted that the above probability of 0.23 is of value within the domain of that particular school district, and even for that particular test; it is devoid of any universal validity. If the same test were given to all tenth graders throughout the nation, for instance, the probability may come out to be quite different from 0.23. But, quite often, it happens that the investigator, in this example, the educator, may, at his own discretion, decide that the school district in which he conducted the survey is quite typical and wrongly project his findings as statistical truths for the entire country.

15.7 Nature of Statistics

Statistics pervades our lives, more often than not, unconsciously. Despite its later arrival than probability, at least in terms of quantities, its growth has been very fast, and its presence is felt more obviously than that of probability. A seventh grader who does not yet measure 5′6″ may give up hope of later making the college basketball team. An athletic 140-pounder in the same grade may hope to shine on the football team. That the French are romantic, the British are conservative, and the Americans love dollars, "everyone knows." Scores of mind-sets and opinions like these are quite common. These, like habits, are the flywheels of our lives, with this difference: these have the support of statistics. My son, a seventh grader, came up with the idea that all French people have big noses. He cited one of his classmates with a big nose who told my son that he was French. I had little success persuading my son that all French people may not have big noses. He remembered yet another French student with a huge nose, who was his classmate two years ago.

The two elements that are conspicuous in such ideas are, firstly, the large numbers implied, like "all French" and "all British," and, secondly, the human tendency to project what is observed on a small scale onto all those that have not been observed but happen to belong to the same class. For example, public-opinion polls and job ratings of presidents, which are fashionable and influencing factors in the United States, are of this kind. Though most people do not know how such "news" came to be, they accept these findings, with due respect, as *statistical truths*. They are derived from a relatively small number of people, compared to the national population, who are considered *representative samples*. Collectively, the process may be called *surveying*, the instruments used being various forms of questionnaires, and the mass of opinions or numbers (as in the case of job ratings) collected being the data.

Another aspect of statistical truth, just as important and relevant to us in the context of experimental research, is derived from a process somewhat the reverse of the above; the common factor is the need to encounter a large number of entities, whose the existence is given and from which the essence, which should represent the entire class, a generalization, needs to be extracted. When such entities are numbers, known as data, they can be subjected to well-known statistical procedures. Collection, tabulation, and presentation of data, often followed by (or following) various methods of analysis, usually of mathematical nature, as a way of describing the data, are known as *descriptive* (or *theoretical*) *statistics*. An instance is describing candidates' performance on a competitive examination relative to age, sex, race, education level, and so forth.

Analysis leading to derivation of inferences (e.g., predicting the outcome of a forthcoming election) and applications of these inferences (e.g., forming a scientific hypotheses or theory) is collectively known as *inferential statistics*. Because the intention of this book is to deal with experimental research involving multiple causes and multiple effects, inferential statistics will be discussed in Chapter 18 on planning experiments.

In this chapter, we will confine our discussion to descriptive statistics, assuming that (1) the decision has been made to select samples, (2) sampling has been done using appropriate methods, depending on the situation, and (3) the required measurements of the samples have been taken, these measurements constituting the data.

With such data on hand, we will look at the computation and significance of two statistical measures: (1) The *central tendency* (*location*), popularly known as the *average*, and (2) the *uniformity* (or the lack of it) among values, known as *dispersion*, often also referred to as the *spread* or *variation*.

15.8 Measures of Central Tendency (Average)

The word "average" is so familiar in common language that it does not need a number (or an algebraic quantity) to specify it. But an average from the viewpoint of statistics is a number (or an algebraic quantity) extracted from a set of numbers (or algebraic quantities), all of which have some common significance. A definite numerical (or algebraic value), the simplicity of extraction and (later) of application, and the representation of every item of a set are some of the desirable attributes of averages. The following are some variations of the average.

15.8.1 Arithmetic Average (Sample Mean)

By far the simplest and most commonly used average, the *arithmetic average* is obtained by adding all the items of the set, numerical or algebraic, and dividing the total by the number of items. For instance, if a runner runs four, five, three, five, four, six, three miles on consecutive days of a week, the arithmetic average, also known as the *sample mean*, is $(4 + 5 + 3 + 5 + 4 + 6 + 3) \div 7 = 4\,{}^{2}/_{7}$ miles per day. Some notable features of this number are as follows: Though extracted from the numbers corresponding to each day's run, it is different from each. It is lower than the highest, and higher than the lowest value. There is no way of retrieving from this the number corresponding to any particular day's run. As for accuracy, though $4\,{}^{2}/_{7}$ can be "improved" as 4·2857, expressing it as $4\,{}^{2}/_{7}$, or even as "about 4π" is quite often adequate. That its accuracy is divorced from that of the individual items from which it is extracted is irrelevant; it is likely that each day's run was not terminated exactly at 3.0000 or 4.0000 miles, and so forth. While its simplicity in derivation is obvious, its simplicity in application is as follows: Suppose for the same runner, the runs in the following week were three, four, three, six, two, five, and two miles. The arithmetic mean for that week is (3 + 4 +3 + 6 +2 +5 + 2) ÷ 7 or 3 $^4/_7$. If it is now desired to get the average for the entire fortnight, it can be obtained as the arithmetic mean of 4 $^2/_7$ and 3 $^4/_7$, namely (7 $^6/_7$ ÷ 2) = 3 $^{13}/_{14}$, which yields the same result as adding all the numbers corresponding to the fourteen days and dividing the total by fourteen.

Generalizing the situation now in mathematical terms and inserting the feature of *sampling,** let x be the variable representing the number of miles run on a given day. Suppose seven days have been *randomly* selected on which the miles run are measured. They can be recorded as $x_1, x_2, x_3, \ldots x_7$. If, instead, we have n (number of) variables of x, symbolized as $x_1, x_2, x_3, \ldots x_n$, each a numerical measurement, then the arithmetic average (or sample mean) denoted by x is given by

$$\bar{x} = \frac{\sum_{i=1}^{n} x_i}{n}$$

The arithmetic mean is both simple and convenient, hence, its popularity. But there are applications where it may lead to wrong inference. For instance, imagine two groups of school children being taken out on a field trip. Individuals in the first group have for their lunch money \$3, \$5, \$4, \$5, \$3, \$4, and \$3, averaging to about \$3.85, and those of the second group have \$20, \$1, \$10, \$20, \$0, \$0, and \$20, averaging to about \$10×14, indicating a more well-to-do group. But when lunchtime arrives, two children of the second group have to go hungry, and one more can afford nothing more than a candy bar, whereas all those in the less-well-to-do first group can afford to have good lunches. The arithmetic mean above, thus, cannot supply the inference needed to know which group can afford lunch for all and which cannot. A few large numbers, even one, may make the average higher without any clue as to how the total is distributed among the members of the set.

15.8.2 Weighted Mean

Here is another case in which the arithmetic mean, at best, leads to wrong inference. Imagine that a department store's clearance-sale advertisement contains statements such as "Reduced prices on many items" and "Prices between only \$4.00 and \$10.00." The bargain hunters, most of them concentrating on \$4.00, and only a few calculative ones, counting on (\$4.0 + \$10.0) ÷ 2, or \$7.00, rush in. With some looking around and a few more than usual pointed questions, one may find out the details listed in Table 15.2.

Table 15.2 *Details behind a misleading advertisement.*

Items marked (\$)	Number of such items marked for clearance
4	1
5	0
6	3
7	4
8	20
9	80
10	200

Even those customers who are calculative are dismayed looking at the large number of items marked over \$7.00. With a little more patience, they calculate what they think the "fair" average price comes to, "fair" meaning "not misleading." Not only the prices, but also the number of items at each price level, should be accounted for. With the large numbers exerting more "weights," this quantity is known as the *weighted mean*, and its value is

$$\frac{1\times 4 + 0\times 5 + 3\times 6 + 4\times 7 + 20\times 8 + 80\times 9 + 200\times 10}{1 + 0 + 3 + 4 + 20 + 80 + 200 = 9.51} = 9.51$$

which is very close to the highest price, \$10.00. Smart shoppers, despite being able to calculate, will be deeply dismayed but learn the lesson that a simple arithmetic mean, by its very simplicity, can mislead.

Generalizing the above example mathematically, if x_1, x_2, x_3, . . . x_n have weights w_1, w_2, w_3, . . . w_n, respectively, then the weighted mean, symbolized by $\overline{x}_w$, is given by

$$\overline{x}_W = \frac{\sum_{i=1}^{n} w_i x_i}{\sum_{i=1}^{n} w_i}$$

As a special case, when all the measurements are weighted equally, with let us say $w_i = 10$, we get

$$\overline{x}_W = \frac{\sum_{i=1}^{n} 10\left(x_i\right)}{\sum_{i=1}^{n} 10} = \frac{10\sum_{i=1}^{n} x_i}{10}$$

which is, by definition, $\overline{x}$. The weighted average, then, is the same as the arithmetic average of the set of measurements.

15.8.3 Median

This is another kind of average for the given set of terms (or measurements). Finding the *median* of a group of numerical (or algebraic) terms consists of four steps:

1. Arranging a series of all the terms in an order of (usually increasing) magnitude
2. Counting the number of terms in the series
3. Locating the middle term, whose magnitude, or numerical or algebraic value, is by definition the median of the group
4. Specifying the middle term as the median

For instance, in an examination taken by a group of candidates, the scores in percentage, entered by a tabulator, are 52, 45, 59, 83, 32, 49, 56, 68, 40, 54, 90, 43, 63, 42, 76, 87, 69, 93, 76,

58, 61, 79, 94, 92, 46, 68, 71, 81, 73, 94, 52, 76, 84, 50, 88, 65, 70, 97, 57, 78, 96, 63, and 61.

Looking at the above data, it is evident that it is presented in no order of any kind. This is often referred to as *ungrouped data*. Whoever made these entries, the tabulator, did his job mechanically, meaning that he did not, or chose not to, see the importance of these data as a way of answering some questions in the future. What is relevant to us is that the logbook of an experimental scientist is very often filled with similar entries, which find their way into it as his experimentation yields results. The difference between this experimenter and the above score tabulator is that the experimenter is all too conscious of the importance of his data, but he chooses to do the analysis at a later time.

Applying the steps mentioned above, on the exam scores data, will impart the following changes:

1. Arranging the numbers in increasing order: 32, 40, 42, 43, 45, 46, 49, 50, 52, 52, 54, 56, 57, 58, 59, 61, 61, 63, 63, 65, 68, 68, 69, 70, 71, 73, 76, 76, 76, 78, 79, 81, 83, 84, 87, 88, 90, 92, 93, 94, 94, 96, and 97
2. Counting the number of terms, of which there are forty-three
3. Locating the middle term, which is the twenty-second (with 21 terms on either side)
4. Specifying the numerical value of the middle term, which is sixty-eight

∴ the median of the set is sixty-eight.

A minor conflict can arise: suppose the group contains an even number of terms. The solution is simple: the mean of the two middle terms is the *median*. If, for example, the above series contained forty-two terms, then the mean of the twenty-first and the twenty-second terms would be taken as the median, with twenty terms below as well as above. The median as an average has some merits. It can be determined easily; its arithmetic or algebraic value is definite; and it is not influenced by extreme values, high or low, and is thus stable. Also, comparing two or more groups using medians as criteria can be justified, particularly when two groups contain large numbers of terms. But there are also some drawbacks. The *combined median* of two or more medians of subgroups cannot be justified; thus, the median has no algebraic property. Another drawback arises when the terms to be averaged are not assigned a one-to-one correspondence with the referents. For instance, in a company, twenty-two employees work eight hours per day, four work ten hours per day, and two work twelve hours per day. The median of the hours, without reference to the corresponding number of employees, is ten hours per day, which is clearly wrong. The right median is eight hours per day because the middle numbers of the total of twenty-eight employees are the fourteenth and the fifteenth, each working eight hours per day.

The median is frequently employed when the attribute to be compared cannot be subjected to quantitative measurement, but nonetheless a rank or gradation is meaningful. The score of the eleventh ranked student in a class of twenty-one students is the median, but his score need not be the mean of those of the twenty-one students.

15.8.4 Mode

In a group of relevant numbers, that particular number that occurs most frequently is known as the *mode*. For instance, in the group representing the scores of forty-three candidates considered before (see Section 15.8.3), the score seventy-six occurs three times, with no other occurring either as many or more times; hence, by definition, seventy-six is the mode. The immediate inference to the examiner is that seventy-six is the "most typical" of the scores. It requires only a quick glance to notice this, which is an advantage. But what if there were also three scores of forty-five in the same group, which in no way is impossible; which one then, seventy-six or forty-five, is to be taken as the mode?

It is often convenient to form, out of a group of numbers, many subgroups of equal intervals, then to select that subgroup with the most members as the mode. The group referred to above, for instance (see Section 15.8.3), can be divided into the following subgroups:

Grades between:						
31–40	41–50	51–60	61–70	71–80	81–90	91–100
Number of candidates:						
2	6	7	9	7	6	6

Here, the advantage of using the mode is immediately apparent. The mode is "61–70." But there is nothing to dictate why subgroups should not be made for grades between 31 and 35, 36 and 40, 41 and 45, and so on, or even 30 and 33, 34 and 37, 38 and 41. The subgroups with the highest number of items may then shift. There is thus the element of arbitrariness possible in using the mode.

On the credit side to mode, we may point out that it is not influenced, unlike arithmetic averages, by extreme values, large or small, and in that way, it is somewhat stable. Against that, we should note that mode is not a function of all the elements of the group, as arithmetic or weighted means are and, in that sense, does not truly represent the group. Yet another disadvantage of mode is that the combined mode of two groups cannot be evaluated by knowing the mode of each of the related groups, as is possible with arithmetic means; as such, it cannot be used in any further steps of algebraic manipulations. In view of some of the disadvantages pointed out, mode is less often used in statistics than its simplicity commends. But an outstanding application suggests itself: manufacturing *one-size-fits-all* items like socks and gloves can be based on modes rather than on the comfort of the individuals who unknowingly buy such items along with the discomfort of wearing them. Application of mode is thus more appropriate with qualitative, rather than quantitative, data. If, for instance, with the population of a given city as a set, such subsets of ethnic difference as white Europeans, Hispanics, African Americans, and Asians are to be formed, then, the mode, namely, the subset with the most elements, is referred to as the *typical* population in contrast to the *average* population of the city.

15.9 Measures of Dispersion

In a group of numbers or algebraic quantities of common significance, there can be, and often is, a considerable variation in the magnitude of the terms. The central tendency among them is indicated by an average, some forms of which we have discussed. Another statistical character that distinguishes a group is the degree of variation in, or the *spread*, of various terms of the group around the central tendency. This latter characteristic is known as *dispersion*, often loosely referred to as *variability* or *distribution*. Alternate methods of quantifying dispersion follow.

15.9.1 Range

The difference in magnitude between the largest and the smallest item in the group is known as the *range*. For instance, in a group of numbers, 32, 11, 16, 29, 7, 19, 36, 18, and 21, the range is given by 36 – 7 = 29. Being easy to compute is perhaps the only advantage of range because it requires only a glance through the series. The disadvantage is that it gives no idea of the distribution; for instance, another group of numbers, 8, 7, 36, and 23, has the same range but a very different distribution. Yet another disadvantage made evident in using range as a statistic is this: the range of heights of men in the United States, for instance, requires unquestionable awareness on the part of the investigator of the tallest man and of the shortest. Even a slight reason to doubt such awareness makes the quantitative measure of range at best questionable and very often useless.

Expressed in general terms, if x_l is the largest in the set of measurements and x_s the smallest, then $(x_l - x_s)$ is called the range of the measurements.

15.9.2 Mean Deviation

A more precise measure of dispersion is obtained by calculating the mean deviation, which is meant to give an idea of how closely the various measurements are bunched around the central location. It consists of the following four steps:

1. Finding the arithmetic mean for the group

2. Subtracting the arithmetic mean from each item of the group
3. Adding all the resulting differences, ignoring the "+" or "–" signs
4. Dividing the sum as found above by the number of terms in the group

For the same group of numbers used to demonstrate range above (see Section 15.9.1), we take the following steps:

1. (32 + 11 + 16 + 29 + 7 + 19 + 36 + 18 + 21) ÷ 9 = 21
2. 32 – 21 = +11, 11 – 21 = –10, 16 – 21 = –5, 29 – 21 = +8, 7 – 21 = –14, 19 – 21 = –2, 36 – 21 = +15, 18 – 21 = –3, 21 – 21 = 0
3. (11 + 10 + 5 + 8 + 14 + 2 + 15 + 3 + 0) = 68
4. 68 ÷ 9 = 7.55

This is the mean deviation, an index of the dispersion or spread among the terms of the group; the lower this value, the lower the dispersion, with the limiting value being zero when there is absolute uniformity among the terms. On the side of higher dispersion, however, there is no limit for the value; as such, this number cannot be used as a scale for measuring dispersion.

In general terms, this may be expressed as follows: given a set of n measurements $x_1, x_2, x_3, \ldots x_n$, the mean deviation is given by

$$\frac{\sum_{i=1}^{n} | x_i - \overline{x} |}{n}$$

where $| x_i - \overline{x} |$ stands for the absolute value, meaning ignoring the negative sign.

It should be noted that there are other methods, besides the arithmetic mean, to measure the average. Correspondingly, following the same procedure as shown above [steps (2) to (4)], but based on other kinds of averages corresponding to step (1) above, we may obtain different kinds of deviations. In broad bracket, all such deviations are called *average deviation.*

15.9.3 Coefficient of Dispersion

A step further, extended on mean (or other kinds of) deviation, with the purpose of showing the extent of deviation or spread with reference to central tendency, is to divide the mean deviation as obtained above by the value of the mean (or any other average) itself. Using the same example as above, it is 7.55 ÷ 21 = 0.3595. This is called the *coefficent of dispersion.* As pointed out before, this fraction tending toward zero indicates lower dispersion, but there are occasions when it can exceed 1.0, with no definite limit for such a tendency, and then it becomes less significant.

15.9.4 Standard Deviation

Standard deviation is by far the most precise and also the most used measure of dispersion. It is known to be appropriate when dealing with quantities obtained from probability considerations. In a sense, it is a modification of mean deviation. Whereas finding the mean deviation required that we ignore the signs of quantities obtained as differences between the individuals and the mean, here we are required to square each such difference, thereby avoiding the negative signs. Consonant with this, standard deviation is obtained by taking the square root of the numerical value that emerges in the place of the mean deviation. The procedure for obtaining the standard deviation, using symbols $\bar{x}$ for the mean and Σx for the sum of all n items of the groups, consists of

i. Finding the arithmetic mean of the numbers:

$$\bar{x} = \frac{\sum_{i=1}^{n} x_i}{n}$$

ii. Subtracting the mean from each item:

$$x_i - \bar{x}$$

iii. Squaring each value of such difference:

$$\left(x_i - \bar{x}\right)^2$$

iv. Adding all the squared values:

$$\sum_{i=1}^{n} \left(x_i - \bar{x}\right)^2$$

v. Dividing the above sum by the number of items:

$$\frac{\sum_{i=1}^{n} \left(x_i - \bar{x}\right)^2}{n}$$

vi. Giving the standard deviation, σ, as

$$\sigma = \sqrt{\frac{\sum_{i=1}^{n} \left(x_i - \bar{x}\right)^2}{n}}$$

It may be noted that a standard deviation is always a positive number, except that it becomes zero only in the event that all items in a group of numbers are absolutely identical. It should also be noted that when the group of numbers is not the whole population but a sample of a possible bigger population, statisticians recommend that we modify the standard deviation into an *estimated population standard deviation* from random samples as

$$s_x = \sqrt{\frac{\sum_{i=1}^{n} \left(x_i - \bar{x}\right)^2}{n-1}}$$

where n is the number of elements in the sample. We may note the implication of this in experimental studies. In view of the arbitrariness of the number of trials or replications and the corresponding data in experimental research, s_x, not σ, is appropriate.

Taking another look at the procedure shown above, we notice that a lot of calculations are made necessary, starting with and involving $\bar{x}$. If $\bar{x}$ comes out as a noninteger, which is very likely, and if there is a large number of x terms, the calculation time becomes quite considerable. Doing an algebraic simplification, as shown below, $\bar{x}$ may be avoided, thereby saving computation time.

$$s_x = \sqrt{\frac{\sum_{i=1}^{n}(x_i - \bar{x})^2}{n-1}}$$

$$s_x^2 = \frac{\sum_{i=1}^{n}(x_i - \bar{x})^2}{n-1} \quad \cdots\cdots\cdots\cdots\text{Squaring } s_x$$

$$= \sum_{i=1}^{n}\left(x_i^2 - 2x_i\bar{x} + \bar{x}^2\right) \div (n-1)$$

$$= \frac{1}{n-1}\left[\sum_{i=1}^{n} x_i^2 - \sum_{i=1}^{n} 2x_i\bar{x} + \sum_{i=1}^{n}\bar{x}^2\right]$$

$$= \frac{1}{n-1}\left[\sum_{i=1}^{n} x_i^2 - 2\bar{x}\sum_{i=1}^{n} x_i + n\bar{x}^2\right] \cdots\cdots\cdots\cdots \bar{x} \text{ is a constant}$$

$$= \frac{1}{n-1}\left[\sum_{i=1}^{n} x_i^2 - 2\left\{\sum_{i=1}^{n} x_i \Big/ n \sum_{i=1}^{n} x_i\right\} + n\left(\sum_{i=1}^{n} x_i \Big/ n\right)^2\right] \cdots\cdots\text{Definition of } \bar{x}$$

$$= \frac{1}{n-1}\left[\sum_{i=1}^{n} x_i^2 - \frac{2}{n}\left(\sum_{i=1}^{n} x_i\right)^2 + \frac{1}{n}\left(\sum_{i=1}^{n} x_i\right)^2\right]$$

$$= \frac{1}{n-1}\left[\sum_{i=1}^{n} x_i^2 - \frac{1}{n}\left(\sum_{i=1}^{n} x_i\right)^2\right]$$

$$= \frac{n}{n}\cdot\frac{1}{n-1}\left[\sum_{i=1}^{n} x_i^2 - \frac{1}{n}\left(\sum_{i=1}^{n} x_i\right)^2\right] \cdots\cdots\cdots\cdots \text{Multiplying by } \frac{n}{n} = 1$$

ii. Subtracting the mean from each item:

$$x_i - \bar{x}$$

iii. Squaring each value of such difference:

$$(x_i - \bar{x})^2$$

iv. Adding all the squared values:

$$\sum_{i=1}^{n} (x_i - \bar{x})^2$$

v. Dividing the above sum by the number of items:

$$\frac{\sum_{i=1}^{n} (x_i - \bar{x})^2}{n}$$

vi. Giving the standard deviation, σ, as

$$\sigma = \sqrt{\frac{\sum_{i=1}^{n} (x_i - \bar{x})^2}{n}}$$

It may be noted that a standard deviation is always a positive number, except that it becomes zero only in the event that all items in a group of numbers are absolutely identical. It should also be noted that when the group of numbers is not the whole population but a sample of a possible bigger population, statisticians recommend that we modify the standard deviation into an *estimated population standard deviation* from random samples as

$$s_x = \sqrt{\frac{\sum_{i=1}^{n} (x_i - \bar{x})^2}{n-1}}$$

where n is the number of elements in the sample. We may note the implication of this in experimental studies. In view of the arbitrariness of the number of trials or replications and the corresponding data in experimental research, s_x, not σ, is appropriate.

Taking another look at the procedure shown above, we notice that a lot of calculations are made necessary, starting with and involving $\overline{x}$. If $\overline{x}$ comes out as a noninteger, which is very likely, and if there is a large number of x terms, the calculation time becomes quite considerable. Doing an algebraic simplification, as shown below, $\overline{x}$ may be avoided, thereby saving computation time.

$$s_x = \sqrt{\frac{\sum_{i=1}^{n}\left(x_i - \overline{x}\right)^2}{n-1}}$$

$$s_x^2 = \frac{\sum_{i=1}^{n}\left(x_i - \overline{x}\right)^2}{n-1} \qquad \cdots\cdots\cdots\cdots\text{Squaring } s_x$$

$$= \sum_{i=1}^{n}\left(x_i^2 - 2x_i\overline{x} + \overline{x}^2\right) \div (n-1)$$

$$= \frac{1}{n-1}\left[\sum_{i=1}^{n} x_i^2 - \sum_{i=1}^{n} 2x_i\overline{x} + \sum_{i=1}^{n} \overline{x}^2\right]$$

$$= \frac{1}{n-1}\left[\sum_{i=1}^{n} x_i^2 - 2\overline{x}\sum_{i=1}^{n} x_i + n\overline{x}^2\right] \qquad \cdots\cdots\cdots\cdots\overline{x}\text{ is a constant}$$

$$= \frac{1}{n-1}\left[\sum_{i=1}^{n} x_i^2 - 2\left\{\sum_{i=1}^{n} x_i \big/ n \sum_{i=1}^{n} x_i\right\} + n\left(\sum_{i=1}^{n} x_i \big/ n\right)^2\right] \qquad \cdots\cdots\text{Definition of } \overline{x}$$

$$= \frac{1}{n-1}\left[\sum_{i=1}^{n} x_i^2 - \frac{2}{n}\left(\sum_{i=1}^{n} x_i\right)^2 + \frac{1}{n}\left(\sum_{i=1}^{n} x_i\right)^2\right]$$

$$= \frac{1}{n-1}\left[\sum_{i=1}^{n} x_i^2 - \frac{1}{n}\left(\sum_{i=1}^{n} x_i\right)^2\right]$$

$$= \frac{n}{n}\cdot\frac{1}{n-1}\left[\sum_{i=1}^{n} x_i^2 - \frac{1}{n}\left(\sum_{i=1}^{n} x_i\right)^2\right] \qquad \cdots\cdots\cdots\cdots\text{Multiplying by } \frac{n}{n} = 1$$

Once the numerical value of S_x^2 is derived, the value of the standard deviation is obtained as the square root of S_x^2.

15.10 Tabular Presentations of Statistical Data

Quite often we find statistical data and their analysis are displayed in tabular form for the simple reason that tables, like rows of books in a well-organized rack, make a better presentation; tables are easier than a list or group of numbers to glance through, compare, and contrast, and it is easier to grasp the intent and purpose of the presentations. What is significant to us is that experimental data, after getting collected, are often massive and disorderly; they need to be presented and analyzed the same way as any statistical data. Table 15.3 structures and presents the unordered data from Section 15.8.3. In the same table, such statistical terms as *frequency distribution*, *cumulative frequency*, and *cumulative percentage* are introduced. The same data is then presented in graphical form.

Table 15.3 *Tabular presentation of data.*

(1) Test Score	(2) Frequency Distribution	(3) Cumulative Frequency a	(4) Cumulative Percentage $\frac{a \times 100}{43}$
32	1	1	2.3
33	0	1	2.3
34	0	1	2.3
35	0	1	2.3
36	0	1	2.3
37	0	1	2.3
38	0	1	2.3
39	0	1	2.3
40	1	2	4.6
41	0	2	4.6

Table 15.3 *Tabular presentation of data. (continued)*

(1) Test Score	(2) Frequency Distribution	(3) Cumulative Frequency a	(4) Cumulative Percentage $\frac{a \times 100}{43}$
42	1	3	6.9
43	1	4	9.3
44	0	4	9.3
45	1	5	11.6
46	1	6	13.9
47	0	6	13.9
48	0	6	13.9
49	1	7	16.2
50	1	8	18.6
51	0	8	18.6
52	2	10	23.2
53	0	10	23.2
54	1	11	25.5
55	0	11	25.5
56	1	12	27.9
57	1	13	30.2
58	1	14	32.5
59	1	15	34.8
60	0	15	34.8
61	2	17	39.5
62	0	17	39.5
63	2	19	44.1
64	0	19	44.1
65	1	20	46.5

Table 15.3 *Tabular presentation of data. (continued)*

(1) Test Score	(2) Frequency Distribution	(3) Cumulative Frequency a	(4) Cumulative Percentage $\frac{a \times 100}{43}$
66	0	20	46.5
67	0	20	46.5
68	2	22	51.1
69	1	23	53.4
70	1	24	55.8
71	1	25	58.1
72	0	25	58.1
73	1	26	60.4
74	0	26	60.4
75	0	26	60.4
76	3	29	67.4
77	0	29	67.4
78	1	30	69.4
79	1	31	72.0
80	0	31	72.0
81	1	32	74.4
82	0	32	76.4
83	1	33	76.7
84	1	34	79.0
85	0	34	79.0
86	0	34	79.0
87	1	35	81.3
88	1	36	83.7
89	0	36	83.7

Table 15.3 *Tabular presentation of data. (continued)*

(1) Test Score	(2) Frequency Distribution	(3) Cumulative Frequency a	(4) Cumulative Percentage $\frac{a \times 100}{43}$
90	1	37	86.0
91	0	37	86.0
92	1	38	88.3
93	1	39	90.6
94	2	41	95.3
95	0	41	95.3
96	1	42	97.6
97	1	43	100

A brief description of the above table is due:

- Column 1: It is noticed from the ungrouped data in Section 15.8.3 that the minimum number in the set (of scores received by the candidates) is thirty-two and the maximum is ninety-seven. All the numbers, in increasing order, are listed, even if there are no scores corresponding to some numbers, such as thirty-three through thirty-nine, forty-one, forty-four, and so forth.
- Column 2: There is one candidate who scored thirty-two, and that shows as "1" in column 2. There are no candidates with scores thirty-three, thirty-four, thirty-five, thirty-six, thirty-seven, thirty-eight, or thirty-nine; a "0" is entered against each. There are two candidates who scored fifty-two; hence, there is a "2" corresponding to that row in column 2, and so on. Now, looking at the relation between column 1 and column 2, we may say that corresponding to each row, column 2 shows the number of entities (namely, the candidates) that have the property (of scoring a particular percentage score) shown in column 1.

Another way of saying this is that column 2 shows how frequently, that is, how many times, the property shown in column 1 is repeated. Because this information is obtained for the entire distribution of the property (namely, the percentage scores), the name *frequency distribution* is assigned to column 2.

- Column 3: With scores thirty-two or less, there is one candidate; that shows as "1" in this column. Because no candidate scored thirty-three, we may also say that "with scores thirty-three or less, there is one candidate," hence, the number "1" still in this column. This statement is true corresponding to scores thirty-four through thirty-eight as well, which is reflected as "1" in this column, corresponding to all these scores. Subsequently, there is one candidate, with score forty, making the statement "with scores forty or less, there are two candidates," hence, the number "2" in this column. Then, there is no candidate with score forty-one, and the corresponding entry in column 3 is 2 + 0 = 2. That is followed by one candidate with score forty-two, and the corresponding entry in column 3 is 2 + 1 = 3. And so on. Thus, corresponding to each row, the numbers in column 3 are obtained by accumulating the numbers that have appeared that far in column 2. Hence, the name *cumulative frequency* is given to this column.
- Column 4: The numbers in this column are obtained based on the fact that forty-three candidates took the examination. One candidate is obviously 1/43 × 100 percent of the number of candidates, for a value of approximately 2.3. Corresponding to any row, let us say with seven in column 3 should obtain 2.3 × 7 = 16.2 (approx.) in column 4. And the interpretation is that 16.2 percent of the total number of candidates scored forty-five or lower in the test. Expressing in percentage, rather than in actual number, is convenient and desirable when the number of items in the set is very large.

15.11 Grouping the Data

Grouping is a convenient means of gradation: consider the academic grades "A," "B," "C," and D used in the United States as against percentage scores used in some other countries. To say that in a particular class, in a given subject, 40 percent are "A" students is a convenient and unambiguous statement. When more information is desired, for instance, that there are 34 percent "B," 21 percent "C," and 5 percent "D" students, forms the complete statistical statement for that class in that subject.

When dealing with very large numbers of entities, for example, the income levels of all employees of General Motors, grouping for statistical purpose is nearly unavoidable. Resorting to grouping, for instance, the (hypothetical) data may be stated as shown in Table 15.4:

Table 15.4 *Grouping large data for quick comprehension.*

Percentage of Employees	Income ($)
1	More than 200,000
15	100,000–200,000
76	50,000–100,000
8	Less than 50,000

This, in a sense, provides the complete statistics. We said "in a sense" because it fails to answer the question, for instance, What percentage of the employees have an income of $70,000 to $120,000? But to the extent that such details are of secondary importance, stratification as above, in way of grouping data, provides a means for quick and easy comprehension.

Experimenters very often need such quick comprehension to evaluate the degree of success of an experiment. For instance, let us say that the forty-three students, whose scores were listed in Section 15.8.2, were given that test after having been subjected to an experimental method of learning Spanish through speaking rather than studying grammar. As pointed out earlier, that data, as presented there, is disorderly and ungrouped. Even after arranging it as in Table 15.3, it still remains ungrouped, whereas the same data can be arranged in several ways, one of which is shown in Table 15.5; in this form, it is referred to as *grouped data*.

Table 15.5 *Grouped data.*

Scores obtained (percentage)	Frequency (No. of candidates)
30 – 34	1
35 – 39	0
40 – 44	3
45 – 49	3
50 – 54	4
55 – 59	4
60 – 64	4
65 – 69	4
70 – 74	3
75 – 79	5
80 – 84	3
85 – 89	2
90 – 94	5
95 – 100	2

It is evident that changing ungrouped into grouped data necessarily involves some loss of data. For instance, from the grouped form above, we cannot know how many candidates obtained score fifty. If the grouping is done, instead, as in Table 15.3, in which the number of candidates is listed against every score, from the minimum thirty-two to the maximum ninety-seven (referred to as *raw data frequency distribution*), then no information is lost. But this advantage is obtained at the cost of having a large number of entries. Incidentally, though both tables present scores obtained against frequency, the tabulation is known more familiarly as *frequency distribution* when there is interval grouping as in Table 15.5; for that reason, it is often referred to as *grouped frequency distribution*.

It is to be noted at this point that the intervals 30 to 34, 35 to 39, 40 to 44, and so on, are all of equal distance, that is, within the range of five numbers. While it is customary and appropriate to have equal intervals throughout, the discretion as to whether the range should be five as here, or ten, or any other number, and correspondingly, whether to group with *class limits* different from those shown here, is left to the experimenter.

15.12 Graphical Presentations of Data

The graphical form of presenting data is even more popular because it is easier to grasp and appreciate than the tabular form. As a prelude to graphical presentations, a tabular organization usually becomes necessary. After organizing in tabular form, the data can be presented in three somewhat related, but distinct, graphical forms, known as the (1) *histogram*, (2) *frequency polygon*, and (3) *cumulative distribution curve*; these are popular but are by no means the only forms. A brief description of each follows.

15.12.1 Histogram

This is perhaps the most popular and, in some respects, the most elegant form of presenting the frequency distribution of a given set of grouped data. Before we proceed further, a few terms relative to this method of presentation and its interpretation need to be explained. This is done below referencing Table 15.5; the same data is later presented in the form of a histogram.

- Each group interval is known as a *class*; for example, scores sixty-five through sixty-nine are a *class*.
- In the above class, sixty-five is known as the *lower class limit* and sixty-nine as the *upper class limit*.
- The range sixty-five to sixty-nine (which is five units, allowing only whole numbers) is called the *class length*. The lengths of all the classes are kept equal.
- The value sixty-seven, which is the midpoint between sixty-five and sixty-nine, is called the *class mark* for this class. Since this class length is five units (an odd number), selection of the class mark is easy. If, instead, the class length were an even number, for instance sixty-five to sixty-eight, then the class mark would be $(65 + 68) \div 2 = 66.5$, not a whole number, hence, a minor inconvenience.

Now we turn to constructing the histogram. When completed, it consists of a series of rectangles, one for each class. On an *x-y* graph, all the class marks are first located, with suitable scale, on the horizontal (*x*) axis. On each class mark then, a vertical line (*y*) is drawn, with suitable scale, to represent the corresponding frequency. Now, each rectangle is drawn such that its base equals its class length, spread on both sides of the class mark, and its height equals the corresponding frequency, already marked. A histogram so constructed from the data presented in Table 15.5 is shown in Figure 15.1.

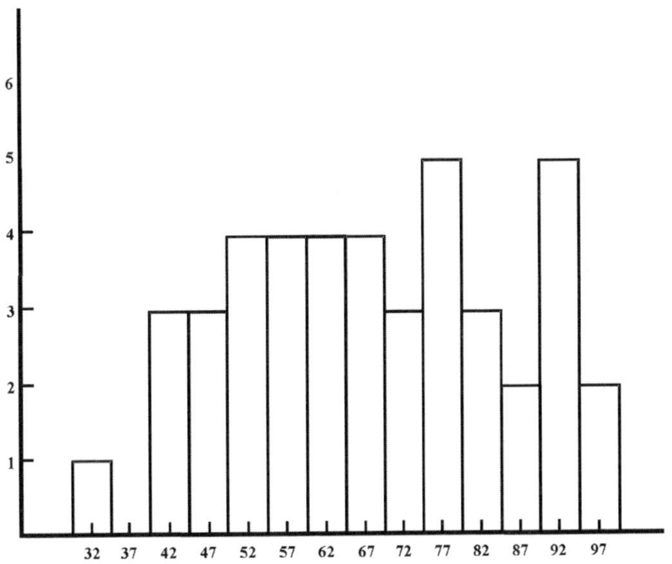

Figure 15.1
Histogram of data from Table 15.5.

In light of the eye-catching visibility of histograms, the following remarks are due:

1. There appears to be a conflict in this graphical presentation relative to the following: The rectangle which shows the number of candidates scoring 50 to 54, for instance, has for its lower class limit 49.5, and for its upper class limit, 54.5. Though we had an understanding that all scores were rounded to whole numbers, we now appear to have reverted to fractional numbers. This happened in the course of translating the tabular data

to graphical form. In the tabular form, there was an entry for scores forty-five to forty-nine, and the next entry was fifty to fifty-four. When this is to be represented in a graph, which needs to have a continuous scale, there will be a gap between forty-nine and fifty. To avoid this gap, the midpoint between these two numbers, namely 49.5, needs to take the place of 50. Once this aspect of apparent conflict is resolved, the other aspects will clear away. Firstly, for the class fifty to fifty-four, the class mark is fifty-two, and the lower and upper class limits should be, respectively, $52 - 5/2 = 49.5$ and $52 + 5/2 = 54.5$. Secondly, in the process of rounding the whole numbers, it is well understood that the rounded value of 49.5 is 50, which belongs to the class 50 to 54; on the other hand, the rounded value of 54.5 is 55, which should belong to the next class, 55 to 59.

2. The area of each rectangle of this histogram is given by: area = base width × height, in which the base width is the class length and the height is the frequency; hence, the area gives the number of candidates in that class. Because each rectangle thus gives the number of candidates in that particular class, the total area of all rectangles together, that is, the area of the histogram in its entirety, gives the value of the total population of the set.

3. The idea of histograms, because of their elegance, is so popular that even when there is no consideration of any class or class length, histograms are used simply to compare different events or entities. For instance, if twelve teachers in a school are to be compared for their teaching effectiveness, each one is used as a class, with no significance of any number. Then, his or her performance, somehow measured and expressed as a number, is plotted as the height of a rectangle. Each rectangle, or a bar, representing one teacher, placed side by side and looking like a histogram, is often called a *bar chart*.

4. With each bar representing one teacher, the bar chart is a popular way of showing how "high" one or two teachers stand out, with reason to be proud, and conversely, how some others stoop, with reason to be humiliated. At this point, whether the method used for measuring teaching effectiveness was beyond any conceivable questionability is conveniently ignored. The precise calculations involved, the statistical numbers that emerged with pristine purity to represent "teaching effectiveness," and the elegance of the bar chart together take the place of "truth."

15.12.2 Frequency Polygon

A variation of the histogram, used less but easier to construct, is the *frequency polygon*. Once the class marks on the horizontal (x) axis and the corresponding frequencies on the vertical (y) axis are marked by points, the frequency points are consecutively connected by straight lines. Further, the two frequency points, one at each end, are connected to the x-axis at a distance of one class length from the extreme class marks.

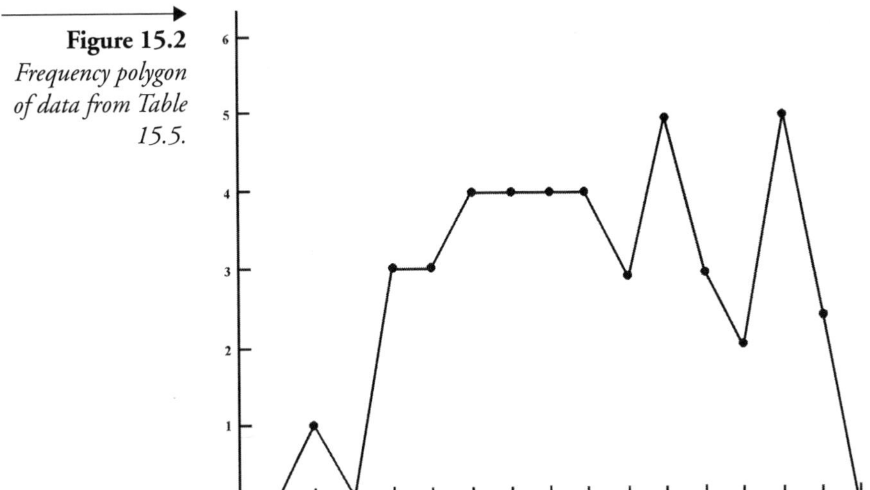

Figure 15.2 *Frequency polygon of data from Table 15.5.*

The completed frequency polygon for the data shown in Table 15.5 (also shown as a histogram in Figure 15.1) is shown in Figure 15.2.

This form of showing distribution is significant in that, whereas histograms are well suited for discrete data like that which we used, a frequency polygon is appropriate for representing data involving continuous change, for example, the time versus the distance of a moving vehicle, in which case, in the limit with large number of data points, would provide a smooth curve of values of y versus values of x.

15.12.3 Cumulative Frequency Distribution

Another form of graphical presentation, which goes a step beyond frequency distribution, is known as *cumulative frequency distribution*. Cumulative frequencies are obtained by simple accumulation of frequencies of classes up to and including each given class in sequence. This is done either from the low values of variables to the high values or vice versa. If from low to high, the upper class limits of all the classes are used as reference points, and if from high to low, the lower class limits are so used. These are marked on the horizontal (x) axis. The corresponding accumulated frequencies are marked as vertical (y) coordinates in the graph. Very often, a slight variation of this, known as *percentage cumulative frequency*, is used; in this variation, the number of frequencies is expressed as a percentage of the total population of the variable. This is handy when the number of entities is very large. Table 15.6 shows the same data as we have used in Table 15.5, now made suitable for graphical presentation of the cumulative distribution curve as well as the percentage cumulative distribution curve, both from low to high values of the variable (score). This data plotted in graphical form is shown in Figure 15.3.

It should be observed that a smooth curve drawn through the points forms an inductive statement derived from the data, more so if there is a large number of points. Using this curve as the reference, it should now be possible to make such deductive statements as, 50 percent of the candidates got scores of sixty-eight or less. But it should be noted that such deduction cannot be extended beyond this particular test and beyond this particular set of candidates.

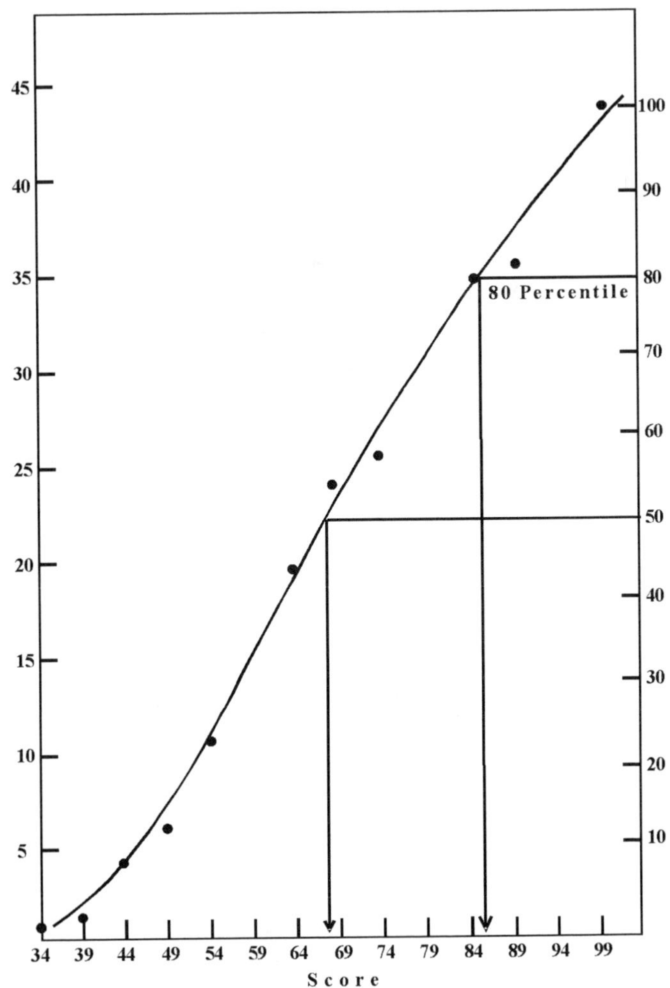

Figure 15.3
Cumulative distribution of data from Table 15.6.

The cumulative percentage curve can also provide the basis for making statements on the relative position of an individual in a set. When a large number of candidates takes a standardized examination, such as a nationwide medical-college admission test, the term *percentile* is often used in reporting the results. If, for instance, a person's score is mentioned to be in the eightieth percentile, the scores of 80 percent of the candidates are lower than his. As tracked in Figure 15.3, such a person should have obtained a score of about 85 percent.

15.13 Normal Distribution Curve

This curve is perhaps the best-known graphical relation in the entire field of probability and statistics. It is said to have been known to eighteenth-century mathematicians Abraham DeMoivre and Pierre Laplace. Now associated with the name of Karl Gauss, the term refers to the distribution of frequencies over a range of a given random, continuous variable. For instance, if the data on the heights of a large number of tenth-grade U.S. boys chosen at random are plotted as a frequency polygon, with very narrow class range (between class limits) like 1/100 inches on the *x*-axis, the resulting figure will be very close to a normal distribution. What may appear surprising, but happens to be a fact, is that such a graphical relation presents itself in many distributions, both natural and man made. The size of oranges harvested in a given orchard, the weight of adult males in a given country, the timings of individuals in a given marathon race, and the closeness to perfection of screws manufactured with a given automatic machine are but a few examples.

Perhaps the most impressive way of seeing the normal distribution curve emerge as a result of pure chance, reflecting probability, is to play with dice. A hypothetical but possible experiment is outlined below:

1. Start with one hundred (or any large number of) identical dice. Shake them together and throw. Some of the dice land with one spot showing up. Also there will be some with two, some with three, some with four, some with five, and the rest with six spots showing up, there being no other possibilities.

2. Count the number of spots showing up on all the dice. This number will be between a minimum of one hundred (if each die lands with one spot up) and a maximum of six hundred (if each die lands with six spots up). The actual number counted can be anything between these two limits. Call that number x_1.

3. Prepare a graph with the horizontal (x) axis ranging from one hundred to six hundred, with a convenient scale. On the graph mark x_1. Thus far, the frequency of x_1 is one.

4. Gather all the dice, shake them together, and throw them again. This time, call the total of all the spots showing up x_2. Mark x_2 on the x-axis. Thus far, the frequency of x_2 is one.

5. Continue operations (1) through (4) a large number of times, say one thousand. Out of these, call the frequencies corresponding to x_1 as y_1 and those corresponding to x_2 as y_2. Also, there will be more values of x, namely x_3, x_4, x_5, . . . x_n, and more corresponding values of y, namely y_3, y_4, y_5, . . . y_n.

6. Prepare the vertical axis of the above graph with a convenient scale to accommodate the highest value among y_1, y_2, y_3, . . . y_n. Plot the graph points corresponding to (x_1,y_1), (x_2,y_2), (x_3,y_3), . . . (x_n,y_n). Connect all the points so plotted.

The shape that emerges will be a close approximation to the *normal distribution curve*, which, indeed, can be obtained by drawing a smooth curve through the plotted points, instead of connecting them with straight lines. The assumption is that all the dice were perfect and that the throws were totally "unbiased." Then, the graphical relation is known to reflect the *probability of pure chance*. The greater the number of dice and the greater the number of throws, both ideally tending toward infinity, the closer the approximation of the resulting curve to that of the normal distribution, emerging from the probability of pure chance. Besides having convenient, mathematically related properties, such a curve serves as a useful approximation to many other distributions, which are less amenable to measurement. Measuring the height of all the tenth-grade boys in the United States and pooling the data, for example, is extremely time-consuming and expensive. If a truly random sample, of say a few hundred, is taken from the entire population, the graphical relation tends to become *normally distributed*, with the expectation that the higher the number of samples, the better the approximation. In exchange for willing-

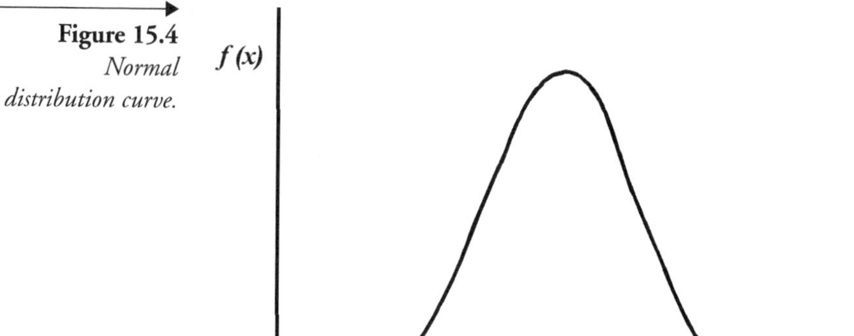

Figure 15.4 *Normal distribution curve.*

ness to accept a small possibility of error, we can thus save an incredible amount of time and expense by measuring the samples instead of trying to measure the entire population. A typical normal distribution curve is shown in Figure 15.4.

And herein lies the key that transforms a bunch of experimental data into a relation, which is covered, end to end, with a single mathematical equation. A frequency polygon, no matter how close the steps of variation of the random variable, is still a collection of empirical data, but once a smooth curve is drawn through the points, which happen to have *Gaussian distribution*, the curve emerges as the theoretical representation of the data with the potential for unlimited application within that domain. The results of experiment have been transformed into theory. Depending on the range of deviation from the mean, and also on the scales used for plotting x and y values, the shape of the curve varies, from squat and wide to tall and thin, but the following properties are common: The curve is variously referred to as a Gaussian distribution, normal probability function, normal probability density distribution, and so on. It is bell shaped and symmetric along a line perpendicular to the x-axis (the random, "independent," "continuous," variable) drawn at the mean of the x values. The mean, the median, and the mode of the values of the random variable coincide. The shape of the curve has the mathematical relation

$$f(x) = \frac{1}{\sigma\sqrt{2\pi}} e^{-(x-\mu)^2/2\sigma^2}$$

where

σ = standard deviation of the random variable x

μ = mean of the random variable x

The continuous, random (dependent) variable $f(x)$ is, obviously, the height of the curve at the particular value of x. The graph actually never touches the x-axis, though it gets closer and closer on both sides to the x-axis as x approaches extreme values. The curve can be interpreted in many different ways and, hence, has acquired many quantitative applications. In nonquantitative, broad terms (perhaps too broad), we may say this: whether natural or manmade, when a measurable quality of individual elements in a given set or population is pooled together, there are increasingly more frequencies concentrated toward the average level of that quality, with fewer and fewer frequencies toward both extremes.

15.14 Frequency Distributions That Are Not Normal

We have noted the preponderance of normal distribution in natural as well as man-made populations. Even so, we cannot take it for granted that whenever there is a set of elements of common significance, the variation of the elements will conform to normal distribution. To do so would amount to abandoning the domain of *probability* and unwittingly accepting *certainty*, for which we seldom can find adequate justification. Indeed, it is not rare that populations are encountered wherein frequency distribution can be distorted and often far removed from being *normal.*

Consider the performance in a competitive examination taken by two kinds of candidates: (1) those who qualify by virtue of their academic achievement, and (2) those who qualify by "reservation" because they belong to an underprivileged segment of the society. For us, the situation is as follows: (1) the population consists of the scores obtained by several hundred candidates, the scores expressed in percentages and in whole numbers; (2) a table of frequency distribution is prepared, considering score as the random variable; (3) using this data, a graph is plotted, with score as abscissa and frequency distribution as ordinates; and (4) connecting these points, a frequency polygon is obtained. It is quite possible, indeed likely, that the frequency polygon registers not one but two peaks. If a curve is now attempted through the terminal points of the ordinates, as is done for obtaining a normal distribution, the same two peaks register again, showing a "deformed" version of the normal distribution. Depending on the number of candidates of each of the two kinds and, of course, the scores received by various candidates, one of the possible curves is shown in Figure 15.5. The second peak corresponding to the higher score, around 80 percent, reflects the performance of the candidates with academic achievement. Such a curve indicates two modes instead of one, in the set of variable and, hence, is referred to as a *bimodal distribution.*

Figure 15.5 *Bimodal distribution.*

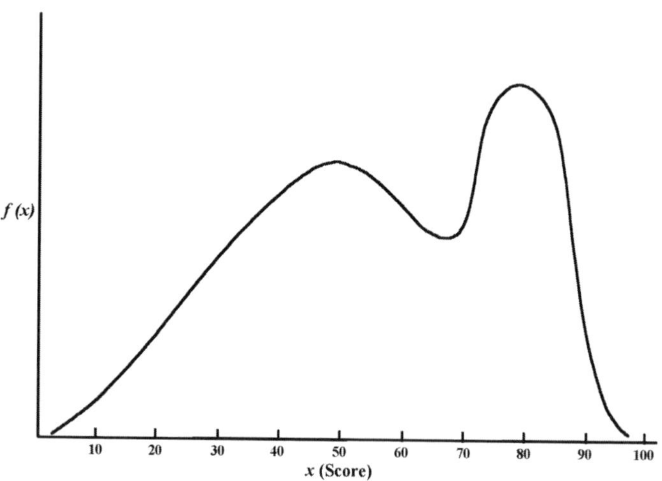

Following this example, some of the other distributions that are departures from the normal are shown in Figure 15.6, with corresponding descriptive names often used. Needless to say, many such curves do not conform to mathematical equations, as does the normal (Gaussian) distribution curve; hence, their utility is rather limited in inferential statistics.

The usefulness of the normal distribution curve lies in its predictive power in various endeavors. In the context of experimental research, it is useful, among other applications, for

- Testing a given hypothesis for its validity
- Sampling experimental materials or parametric variations

Figure 15.6 *Departures from normal distribution.*

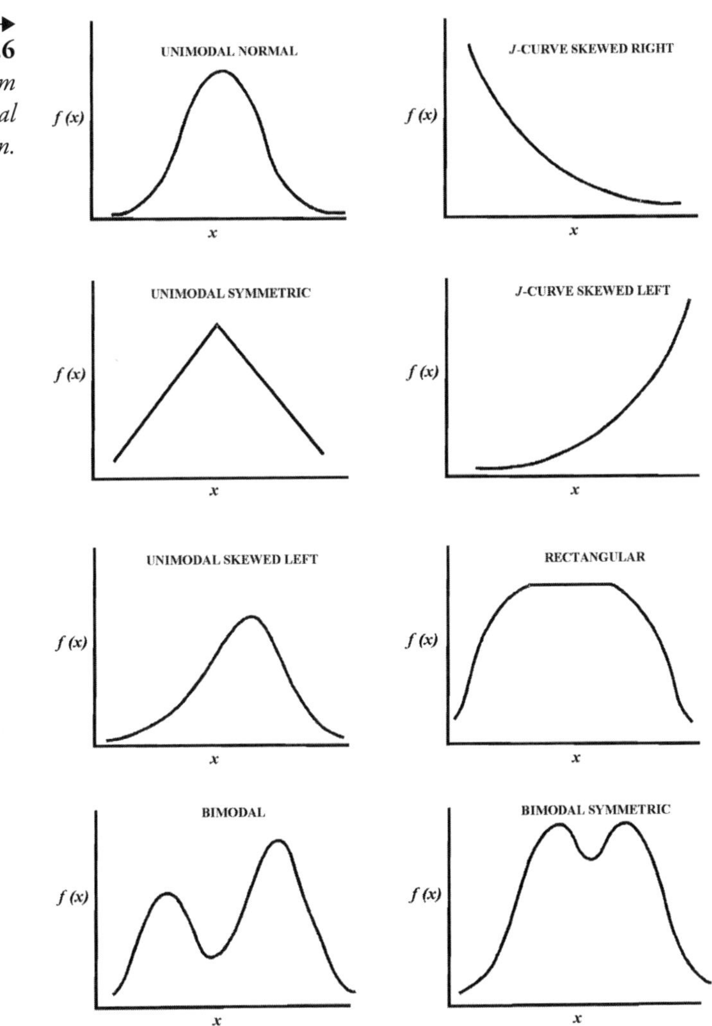

- Testing for the significance of given variables
- Correlating experimental data in an effort to arrive at a law relating the variables

To these, we will return later in the discussion of inferential statistics and while doing the design and analysis of experiments.

15.15 References

1. Bertrand Russell, *The Problems of Philosophy* (Oxford University Press, 1948), 61–62.

2. S. Goldberg, *Probability: An Introduction* (Upper Saddle River, NJ: Prentice Hall, 1960).

3. M. Richardson, *Fundamentals of Mathematics* (New York, NY: The Macmillan Co., 1941), 369.

4. Russell, *Problems*, 139–40.

15.16 Bibliography

Freund, J. E. *Statistics: A First Course.* Upper Saddle River, NJ: Prentice Hall, 1970.

Steen, F. *Elements of Probability and Mathematical Statistics.* London: Duxbury Press, 1982.

Lindgren, B. W., and G. W. McElrath. *Introduction to Probability and Statistics*. 3rd ed. New York: Macmillan Co. 1969.

16

Randomization, Replication, and Sampling

[G]iven a finite collection of facts which has any determinate constitution whatever . . . it remains true that we can, with probability, . . . judge the constitution of the whole collection by the constitution of parts which are "fair samples" of that whole, even when the collection is very large and the samples are comparatively small.

—*Josiah Royce*

Experimentally found relations between independent and dependent variables are not absolute but statistical truths. Nonetheless, they are expected to stand the test of time, in that they are expected to be reproducible any time in future and anywhere else under similar experimental conditions. There are three basic means available to the experimenter to ensure that this happens. The first one is *randomization*. Some typical instances of randomization are described in this chapter. Repeating experiments with the same or purposely varied independent variables, the second means, known as *replication*, is indicated. A statistically justifiable method of sample collection is yet another means to the same end: *reproducibility*. A collection of elements (in our context, samples), known as a *set*, calls for an outline description of set theory. That is followed by a review of some of the more common methods of sample collection.

16.1 Need for Randomization

We may recall that in Chapter 7, discussing the experiment on the benefit of a hypothetical plant food, and also in other contexts, we mentioned the word "random" quite a few times. We may now ask, Why random? Why not pick the first forty or so plants of one kind that the experimenter came across and use the first twenty as subjects and the rest as controls? There are several answers to these questions: Firstly, there is the dichotomy of "subjective" and "objective" in an investigation. What distinguishes scientific investigations is that they are meant to be objective, meaning that an individual's findings as a result of his investigation are not tainted by his prejudice, biases, and wishful thinking, which are subjective elements. As we pointed out before, every inference, every conclusion that is offered, is yet another "brick" in the structure of science and is open for anyone interested to believe or to doubt, to accept as given, or to check for its truth-value before acceptance. A particular experimenter may choose to accept that object or phenomenon or situation which is favorable to his wishful thinking and reject those that are not. An ideal, as a code of conduct that a scientist is expected to strive for, is to free himself by deliberately practicing randomness at all levels, thus giving equal chance to those factors that are favorable to his wishful thinking and to those that are not.

Another circumstance which gives credence to randomization is the fact that nature is filled with variety, with possibly no two "similar" things or events being "exactly similar." Add to this the fact that an experimenter is limited in time and place, whereas the inference or conclusion drawn from his experiments is meant to be beyond the limitation of time and place. Hence, the best he can do is to have "representatives" to experiment on, from different places and, if possible, from different times, and to create an assembly of such representatives; such an assembly to him becomes his "universe." When he finds the truth-value that he surmises, within his limited universe, he dares to project it as good for the outside, real world. This truth has now to contend with the world, in which variety is more a rule than an exception. It cannot be expected to be a "perfect fit" in any specific domain, but it can be a fairly close fit in most domains. Experimentally derived truth is thus essentially statistical in nature; it cannot be absolute. The way the experimenter collects representatives, often called *sampling*, is a reflection of the situation that the truth derived as the end from these representatives serving as the mean, is expected to be a close fit, with various degrees of closeness, in all the domains represented in the sample.

A simple example may illustrate the point. Suppose an experimenter is called upon to find the grain-size distribution of sand in a two-mile-long sea beach. He should not take a bucket full of sand at any convenient location on the beach and proceed to the lab for testing. He needs to collect "representative" samples from different locations over the entire two-mile stretch. Further, he cannot say to himself, Every one hundred yards I will pick a handful at the surface to be representative. He needs some sand samples from the surface, some from, let us say, a foot below the surface, some from flat surfaces, some from crevices, some dry and away from the waterfront, some wet and near the waterfront, and so forth, and to cap it off, the samples need to be collected at "random," that is, not in conformance with any order. The necessity of this last criterion is twofold. First, the experimenter is likely to use any order as a tool for "fixing" the sampling, and thereby "fixing the truth." Second, the size analysis he would come up with may not be found anywhere in the entire beach, but it may be fairly close to the analysis found at various locations with different degrees of closeness. Those locations where the closeness is very high, as well as those where the difference is very high, cannot be predicted. They are likely to be spread along the beach "at random." The nature of the results the experimenter is going to derive is reflected in the way he collects the samples.

Yet another benefit attributed to randomization is that, when used in combination with *replication*, the effects of *nuisance factors*, particularly when such factors are difficult to detect, can be minimized, though not nullified. This obviously rests on the idea that randomization necessarily scatters, instead of focusing, the intensity of effects caused by nuisance factors. For instance, consider a paired comparison experiment in which, let us say, twenty pairs of plants are being tested. If all the *control* plants (A, B, C, . . .) are placed along the south edge of the fenced plot, and all the *subjects* (A^1, B^1, C^1, . . .) are placed along the north edge, factors such as sunshine, wind current, and shadow coverage, which are known to influence plant growth and yield, but are not controllable, may be different on the south edge from those on the north edge. When it comes to comparison of *controls* as a group against *subjects* as a group, there is likely to be a bias unintentionally introduced in favor of one group. Randomization is the remedy to such defect. If plants were placed either at the south edge or the north edge, as decided by the toss of a coin, regardless of whether the particular plant was a *subject* or a *control*, the effect of the uncontrollable factors would be likely to even out.

16.2 Applications of Randomization

Obtaining samples or specimens from a *group* (also known as a *lot* or *population*) of any kind is an activity wherein the application of randomness is of utmost significance. In the case of the study by pairing, a toss of the coin serves the purpose. Randomization has applications beyond sampling, for example, in the sequencing of experiments. We will discuss only two obvious situations. Let us say that there is a hypothesis on hand that the performance of a certain machine is influenced by the temperature of the room. The experimenter wants to test the performance at 50°F, 60°F, 70°F, 80°F, and 90°F, and there should be five replications at each temperature. The usual "commonsense" approach is to set the room temperature at 50°F and to test the performance five different times with adequate time gaps between consecutive tests, then to raise the temperature setting to 60°F, repeat the five tests, and so on, until all twenty-five tests are completed in as many days as required. In this scheme, there is no thought given to randomization. When it is given serious consideration and some randomization is imposed, the scheme of experiment is somewhat like that shown in Table 16.1.

Table 16.1 *Randomization imposed on 25 replications.*

Consecutive Time Period within One Day	Performance Tested for, at Temperatures Shown				
	1st day	2nd day	3rd day	4th day	5th day
1	70	90	50	60	80
2	90	60	80	70	50
3	50	80	60	90	70
4	80	70	90	50	60
5	60	50	70	80	90

From the results—let us say, various readings and calculation for "performance"—of experiments per this scheme, obviously the performance at 70°F is "averaged" based on the five performances at that temperature obtained on different days, at different time periods. But how were the sequences arranged for each day? In the above table, they were "predetermined." To adhere to the idea of randomization, that ought to be avoided. For instance, the experimenter can use five sheets of paper with one of five temperatures (50, 60, 70, 80, 90) on each sheet, then fold the sheets to make the numbers invisible. At the beginning of each day's experiment, he has one of the five folded sheets picked by a colleague who is not involved with the experiment. Whatever temperature is found on the sheet is the temperature at which the first performance test of the day is to be conducted. He then has another folded sheet picked out of the remaining four, the temperature written on that sheet being the second test temperature in sequence for the day. Then, the third and the fourth are picked, leaving the last, the fifth. This random selection determines the time sequence of the experiments for the day. The experimenter does this sequencing process each day, thereby implementing randomization.

Another situation where randomization is usually ignored, but can be implemented, follows: Suppose an experimenter's interest is to test a new synthetic sand binder on the market. He wants to mix 1, 2, 3, 4, 5, 6, and 7 percent by weight of the binder with a standardized sample of sand and test it for strength per standard procedure. Let us say he prepared a plot of strength versus percentage of the binder added; it looks like Figure 16.1.

The numbers next to the curve show the time sequence in which the individual tests were performed. Obviously, in this experiment there is no thought of randomization. If, instead, randomization was imposed somewhat on the same basis as we discussed in the previous case, the sequence could be different, as shown in Figure 16.2.

Obviously, there are several such sequences possible, this one arbitrarily "fixed" to be random. In this example, again, the question may be asked, Though the order 4, 1, 7, 2, 6, 3, 5 does not follow any conceivable mathematical series, is not the very process of predetermining (or fixing) the sequence against the idea of randomization? Since the answer is yes, this defect can be reme-

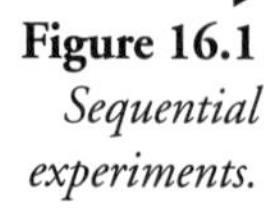

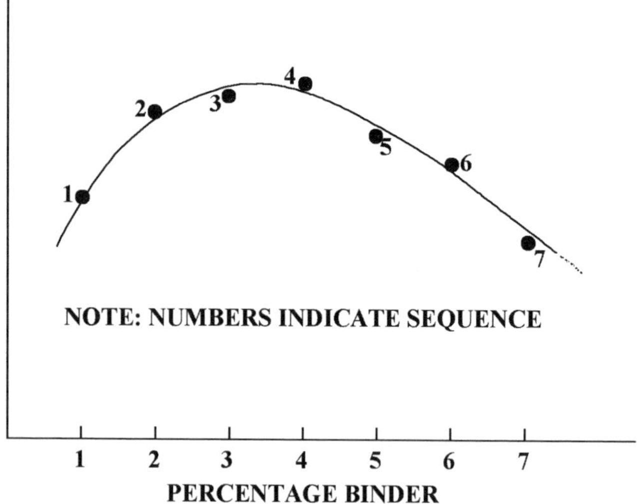

Figure 16.1 *Sequential experiments.*

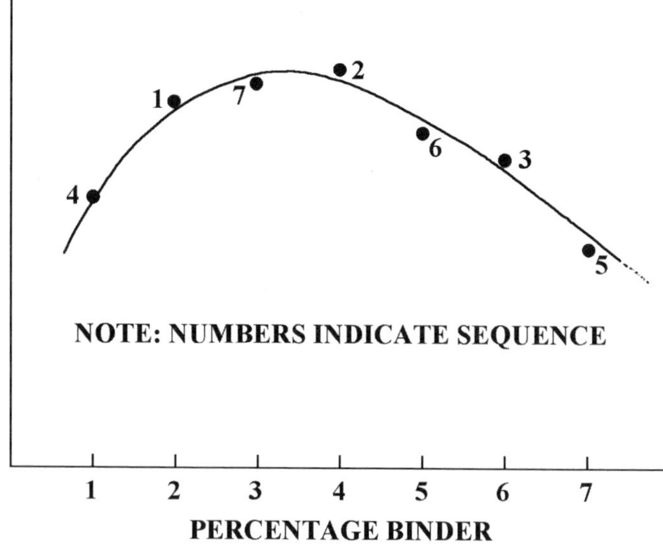

Figure 16.2 *Randomized experiments.*

died as we described previously, by picking numbered sheets that are folded so the numbers are hidden. In this case, each paper would have one number of 1, 2, 3, 4, 5, 6, or 7 (percentages of binder), with the sequence in which they were picked deciding the corresponding sequence for the experiments.

Randomization of land plots for experiments in agriculture, often referred to as *two-dimensional spatial design* or *block design*, is one of the earliest applications, pioneered by R. A. Fisher (1890–1962). Suppose five varieties of a farm crop, say rice, are to be tested for productivity or yield in an apparently similar set of conditions: soil, climate, and agricultural treatment. The total experimental area is not simply divided into five parts, and each variety of crop assigned to one part. Instead, the following arrangement, one of many possibilities, is acceptable. The experimental area is divided into four rectangular "blocks" about equal in size. Each block, in turn, is divided into five "plots" running from end to end of the block, lying side by side, making, in all, twenty plots. In each of the four blocks, the five plots are assigned one to each variety of the crop under the test; this last assignment is done "at random." The method for randomization can be picking the papers with hidden numbers, matching any one of 1, 2, 3, 4, or 5 plots with any one of, say, *A*, *B*, *C*, *D*, or *E* varieties of crop, or by any other methods available. One of the many possibilities of such randomization is shown in Figure 16.3.

Figure 16.3 *Randomized block design for agriculture.*

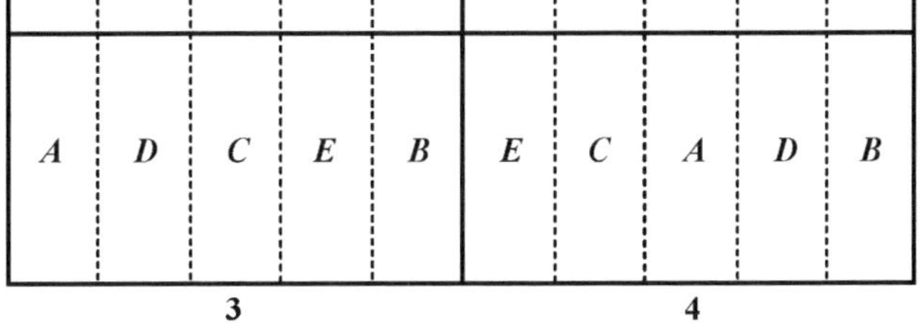

NOTE: ***A, B, C, D, E*** **ARE VARIETIES OF RICE**

	1					2			
C	*E*	*B*	*A*	*D*	*B*	*A*	*D*	*C*	*E*
A	*D*	*C*	*E*	*B*	*E*	*C*	*A*	*D*	*B*
	3					4			

Passing over many more contexts and methods of randomization, we may conclude this list with the case of *opinion polls*, all too frequently used by the media and eagerly accepted by the public. Seldom do we hear how many people contributed to the opinion, through what means of communication, and using what criteria and method for sampling. Needless to say, if the "poll" is "national," it should represent all kinds of varieties in the nation: city, town, and rural; upper, middle, and lower class; learned, educated, and uneducated; men, women, and children; professional, wage earning, and welfare class; old, mature, and young; and so on. Above all these, the important criterion is randomization. How it was implemented, we hardly get to know.

16.3 Methods of Randomization

Throwing the coin and wagering heads or tails is the simplest random process. If there are more than two possibilities, obviously, throwing the coin is useless. Then, we may resort to the method of picking paper strips with hidden numbers, this being suitable for any number including two. In the hope of reducing the subjective part and enhancing randomization, statisticians have devised a few more methods, often elaborate. Two such methods are outlined, using as examples the process of randomizing land plots for agriculture that we have dealt with earlier.

1. **Using a pack of cards.** We have five varieties of rice to be tested. Number those as varieties 1, 2, 3, 4, and 5. We have four blocks, each with five plots, together making twenty plots. We take a pack of cards numbered one to one hundred (one hundred is divisible by five), shuffle thoroughly, and lay them in random order in a line. Take the first card; let us say it bears the number seventy-two; this number, when divided by five, leaves a remainder of two. Assign the first plot, starting from one end, with rice variety 2. Then, pick the second card, let us say this bears the number thirty-nine; this divided by five leaves a remainder of four. Assign the second plot with variety 4; and so on. When the remainder is zero, it corresponds to variety 5. If there are repetitions within a block (of five plots), discard that card, and go to the next. By this method, proceed until all twenty plots are assigned, each with a variety of rice. Suppose the number of varieties was six instead of five. Then, we would need to use ninety-six cards, not one hundred, as ninety-six is divisible by six.

2. **Using tables of random numbers.** Known as *random digits* or *random permutations*, such tables are available in print, attached as appendixes to many books on statistics; a typical one is shown in Table 16.2. In such a table, we may start anywhere, choose to go on any line, horizontally, vertically, or diagonally, and use each number instead of a numbered card, as in the previous method. If there is such a table with each number of four digits instead, we may choose the last two digits each time, or the first two, or the middle two, and so forth. If any number yields unwanted repetition, or is superfluous or unsuitable (example, four to be divided by five), we simply ignore it and go to the next number. Since randomization is the aim, freedom in using these tables is practically unlimited. The advantage of this method is that instead of a pack of numbered cards, and the need to shuffle and lay the cards in a random order, all we need is a printed sheet or to open a book to the page with such a table.

Table 16.2 *Typical Random Numbers*

80125	84984	47804	90908	11106	43995	83182	39647	41504	23286
69743	16618	59364	73113	89143	89412	39331	84989	92091	01383
41686	85487	91526	28591	54635	77149	59117	30291	68266	65760
80967	75537	81999	02257	62232	00956	24954	88932	90733	35767
24148	95247	12340	39407	71596	53961	61089	68699	38891	87153
22989	64262	12716	32910	32303	18783	65166	56622	93342	14032
54147	01638	95954	66666	30544	67089	04524	19251	57440	69100
07529	10668	23743	02743	10252	47893	83969	54252	47327	31685
36379	13588	44587	31015	34971	25146	33188	05218	17157	65663
38653	73761	61363	95667	03372	35800	58711	15872	33342	02963
72327	65811	53782	01608	38741	58353	51594	48982	85028	75444
41133	06312	13340	18870	27204	83187	91970	91498	17234	52283
15039	81095	50787	28452	61100	39538	25225	92624	18517	77361
40499	67587	16761	25929	43836	43466	80409	95407	46777	22668
46910	72907	18515	12710	11580	52823	95769	71506	39644	66877

Table 16.2 *Typical Random Numbers (continued)*

15009	81751	07942	14046	54993	68001	13782	91933	61130	61752
10538	10295	62995	16527	55334	05736	92168	35393	01026	26984
70204	91225	78307	55577	78715	54507	21486	26920	52995	98095
30403	98849	55318	99947	23625	74643	85157	45893	49287	03567
75147	49930	47054	08485	91397	25614	83669	08353	61573	49629
20770	53498	05412	19184	25997	06100	27128	43137	77812	13101
51096	90416	18721	42390	31517	28366	30073	89021	40881	36162
05027	43924	37581	31418	57010	05808	75544	68156	75440	64496
22013	11299	76690	92730	10867	12748	58655	44844	11933	16752
50232	30821	45382	85723	15635	85910	19874	61262	74598	41321
20724	99075	91270	13936	74962	15346	05181	52254	42138	18237
65692	61084	48856	34766	09098	87381	29763	65051	91174	80750
81215	08824	06387	10900	83463	19773	83029	81689	66067	38729
86127	96878	53819	10715	67213	53160	17249	44596	76354	33601
26483	16992	89421	15216	71632	83429	96263	16342	15595	48978

(Source: John Wiley & Sons Inc. 1964; *Statistics and Experimental Design: In Engineering and the Physical Sciences*, Vol. 1; Norman L. Johnson- Fred C. Leone)

Though the previous two examples were mentioned in reference to randomizing the agriculture plots, needless to say, these methods can be adapted to many other contexts, the essence being to match members of one category at random with those of another.

16.4 Meaning of Randomization

After dealing with these various contexts—there are many more—of randomization, we may ask, What is really meant by randomization? The closest we can get to the meaning of "random" is "done without previous calculation" or "left to chance." Any attempt to define randomization is known to be philosophically hopeless for it begs the idea of probability. Is zero *probability* equivalent to 100 percent *randomness*? Is absolute randomness practical, even possible? Is randomization quantifiable? This is a subject area rife with questions and opinions. In the meanwhile, an experimenter can only impose randomness deliberately in all conceivably required contexts, to an extent possible, knowing that there may not be anything like 100 percent randomization.

16.5 Replication

We have already used this word, even before defining it, as we shall do now. *Replication* is repetition with some difference. When we speak of repetition, we normally imply a sequence in temporal order; what we do now, we may do after a lapse of time, no matter how long. This is replication of the simplest kind. When repetition is concurrent or simultaneous, that is replication too. For instance, in the agricultural experiments we have discussed, in which different varieties of rice were tested for yield, we find in the experimental setup that each variety of rice was cultivated not in only one, but in four different plots, under identical conditions, but physically separated. The yield from each plot can be compared with that from any one or all of the other plots. But, as far as the yield of that particular variety is concerned, the statistical average of yield from all four plots is relevant, and for this purpose, the experimental details are equivalent to four repetitions, though done simultaneously.

Another extension of replication is the case of paired comparison experiments we discussed in Chapter 7, for testing a plant food, where A and A^1 were two flowering plants of one kind, B and B^1 of the second, C and C^1 of the third, and so on. They were experimented on under identical conditions. All these tests of ten comparisons were virtually ten replications in the sense that they were all one experiment testing the benefit of one plant food on one class, namely flowering plants, though this class contained ten varieties. Here we had identical conditions relative to the treatment, but variety relative to the recipient of the treatment. The intention of the experiment was to form a generalization of the benefit of a particular plant food on flowering plants. It was one experiment with ten replications.

In the generalization of the one-cause-one-effect, x–y relation, the effect on the dependent variable, y, of changing the independent variable, x (usually a quantity), is the essence of the experimentation. The hardware part of the experimental setup, the calibration, the measuring devices, and so forth, remains unaltered. But for the change in parameters, experiments conducted are virtually a series of replications. Replication with parametric varieties, or simply *parametric replications*, may be an apt name for these.

Now, what is the purpose of replication? Common to all three variations we discussed, we may say that replication increases the confidence of the experimenter in his inferences. Whether this can be quantified into a *confidence factor*, we will discuss in Chapter 19, though we may say at this point that the more the *replication*, the more the *confidence*, with the highest confidence still falling short of certainty, just like the highest number is still not infinity. The relevant, practical question is not how *many*, but how *few*, replications are necessary. For the purpose of obtaining sufficient confidence to make a statement or attempt a generalization, what should be the minimum level of confidence? Statisticians have worked out the answer to this question in various situations relative to availability of data; we shall deal with these in Chapter 19.

16.6 Samples and Sampling

In Chapter 15 we dealt with the statistical properties of an arbitrarily selected group of numbers, which can be considered a *set*. Every *element* of the set, in most cases, contributes toward determining the required property. If the population of the set is too large to handle, or if economy in terms of time or money is crucial, it is normal practice in statistical studies to randomly select a smaller number of elements, forming a *subset*, known as a *sample*, and to predict or infer the statistical properties, such as average and deviation, of the whole population from those of the sample. The inferred properties thus obtained cannot be expected to give the corresponding property of the whole population with 100 percent accuracy for the simple reason that all the elements of the population are not allowed to participate in deriving such property. But such properties derived from the sample, known as *sample information*, will bear probabilistic relations to the corresponding properties of the entire population. This explicit combination of probability and descriptive statistics is referred to as the *Theory of Statistical Inference*; some applications of this we will see in Chapter 19.

Because our context is experimental research, we may need to encounter a number of items within a class in various forms: experimental specimens, different parameters as causes in an experiment, the steps of variation of a given parameter, the combination of many parameters at many levels as causes, the readings of measuring instruments used to record the effects in an experiment, the enumeration of yes-or-no results as effects of an inquiry, and so on. In all such cases of laboratory experiments, the numbers involved are not necessarily large.

There are disciplines, however, in which the experiments are not confined to the laboratory, for example, psychology, education, and sports, in which *surveys* of various sizes are used, hence, in which large numbers of elements in a given class are fairly common. Sampling in such cases is a very important component of experimentation. Sampling has several contextual variations, for instance:

1. Collecting data, such as height from a small number of male tenth graders to represent the data of all male tenth graders in the country
2. Selecting a small number of plants to represent all plants of that kind in a greenhouse, the purpose being to test the effect of a plant food (see Chapter 7)
3. Selecting a small number of manufactured items issued from an assembly line, the purpose being to test these items for performance as part of quality control

In the context of experimental research, sampling is often unavoidable, even in principle. For instance, the experiment on the benefit of a plant food discussed in Chapter 7 involved such questions as

1. Is the new plant food beneficial to the plants?
2. If yes, to what extent is it beneficial?
3. Are there any deleterious effect on the plants?

To answer such questions, it is reasonable that only a limited number of plants, a small percentage of the whole lot, be subjected to the test. When a new drug is to be tested on humans as a possible remedy for a specific disease, the situation is even more critical. Sampling in such circumstances is not only unavoidable, but it is even necessary because it embodies part of the hypothesis. The basis of all sampling is that out of an available *set* (also known as a *lot*, *group*, or *population*) of items, a smaller set needs to be selected. The selection should be done in such a way that *every member of the original set has an equal chance of becoming a member of the smaller set*. When this criterion is met, the procedure followed is known as *random sampling*. Thus, the idea or notion of a set plays an important role in sampling. The logical elaboration of this, expressed in mathematical terms, is known as *set theory*. A brief introduction to this follows.

16.7 Notions of Set

- *Set*: is a collection of elements
    - Capital letters (A, B, C . . .) are customarily used to symbolize sets.
- *Elements*: may be numbers, measurements, materials of any kind, plants, animals, men, or any other objects or entities.
    - Lowercase letters (a, b, c, . . .) are customarily used to symbolize elements.
- $X = \{a, b, c\}$: This is the way of denoting the set X.
    - Elements of a set are enclosed by braces, with commas in between.

  Also
- $X = \{c, a, b\}$: The order of elements is not relevant.

  Sets of Numbers:

Example 1:

$S = \{1, 3, 5, 7\}$: S is the set of positive, odd integers up to (and including) seven.

Example 2:

$M = \{1, 2, 3, 4, \ldots, 50\}$: M is the set of all positive integers up to fifty.

Example 3:

$N = \{2, 4, 6, \ldots, 24\}$:

N is the set of positive, even integers up to twenty-four.

Example 4:

$Y = \{5, 10, 15, 20, \ldots\}$:

Y is the set of all positive integers, which are multiples of five.

- When numbers are involved in a series, it is customary and desirable to list a sufficient number of elements, in a familiar order, so that the relation among numbers can be readily noticed.

- *Finite set*:
  has limited numbers of elements, as X, S, M, and N above.
- *Infinite set*:
  has an unlimited number of elements, as Y above.
- $n(X)$:
  is the way of denoting the number of elements in set X (which is 3, see above).

Example:

$n(X) = 3$:
is the symbol for saying that set X has three elements.

Also

$n(S) = 4$:(see above).

• *Null (or) empty set*:
is a set with no elements, symbolized by ϕ.

Example:

$H = \phi$:
H is the set of horses that have horns.

Then

$n(H) = 0$

Universal set:
is the largest set of a certain kind of elements, symbolized by U.

Example: All those who have visited Disneyland.

Subsets: All other sets are subsets of U.

Examples:

All children who have visited Disneyland in 1999.

All newlyweds who have visited Disneyland so far.

- $A = B$
  means that set A and set B contain exactly the same elements.

  Example:

  $A = \{l, m, n, p\}$; $B = \{n, m, p, l\}$

- $A \neq B$
  means that sets of A and B do not contain exactly the same elements.

  Example:

  $A = \{s, r, i, r\}$; $B = \{s, l, i, n\}$

- Set A and set B are mutually *exclusive* (or *disjoint*) if they contain no common elements.

  Example:

  $A = \{v, 1, s, h, w\}$; $B = \{p, r, n, k\}$

- $A \subset B$
  means that set A is a subset of set B.

  Every element of set A is in set B.

  Example:

  $A = \{3, 4, 6, 7\}$; $B = \{3, 4, 6, 7, 12\}$

  Every set is a subset of itself.

The null set is a subset of every set.

- A' is the complement of set A.

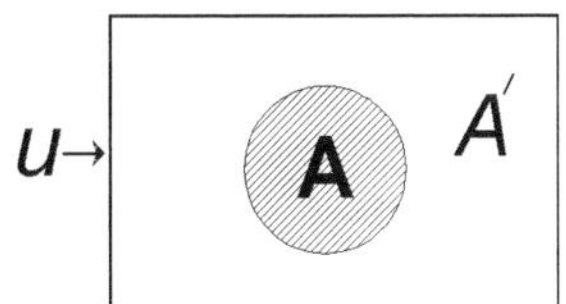

A' is the set of all elements in the universal set, except the elements in A.

Example: $U = \{1, 2, 3, 4, 6, 9\}$; $A = \{2, 4, 9\}$; $A' = \{1, 3, 6\}$

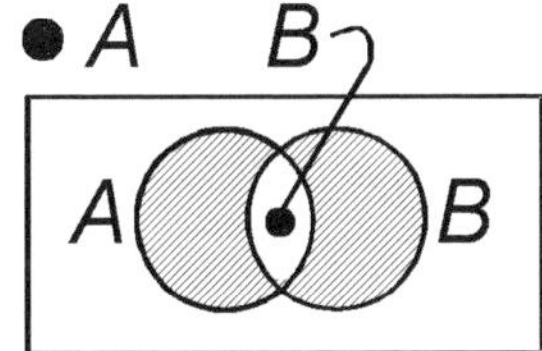

Symbol for intersection of sets A and B

Set of all elements that belong to both A and B

Example: $A \cap B = \{a, f, g, l\}$; $A = \{a, d, f, g, l\}$; $B = \{a, c, f, g, l\}$

- $A \cup B$

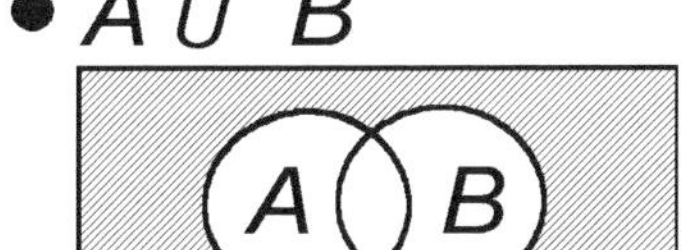

Symbol for the union of sets A and B

Set of all elements that belong to at least one of the two sets A and B

Example: $A \cup B = \{a, b, c, d, e, g, h\}$; $A = \{b, c, e, g\}$; $B = \{a, d, h\}$

16.8 Permutations and Combinations

After doing sampling, the result of which is the set of a smaller number of elements, the relevant questions are

1. In how many different ways can we arrange the given (small) number of elements of the set, dealing with all of them together?
2. How many different subsets can we form out of a given set of elements, conforming to certain specified constraints?

These questions lead us to what are known as permutations and combinations in a set.

With the prospect of having to deal with experiments that have many causes acting together, it is necessary to know the permutations and combinations possible within a set of a given number of things, objects or otherwise. Any arrangement of a group or set of specific things in a specified order (*AB* is different from *BA*) is a *permutation*, whereas a group or set of specific things, disregarding order, is a *combination* ($AB \equiv BA$).

16.8.1 Permutations

Consider a set of three books, marked *A*, *B*, and *C*, to be placed in a rack that has room for only two of these. The permutation of these three books, two books taken at a time, are six in number:

AB	AC	BC
BA	*CA*	*CB*

Further, consider a set of four books, marked *A*, *B*, *C*, and *D*. Permutations of these when three books are to be taken at a time are twenty-four in number:

ABC	BCD	CDA	DAB
BCA	CDB	DAC	ABD
CAB	DBC	ACD	BDA
CBA	DCB	ADC	BAD
ACB	BDC	CAD	DBA
BAC	CBD	DCA	ADB

The number of permutations can be rationalized without having to write them as above. We take a close look at the first case: three books, taken two at a time. In the first position, any one of the three books can be placed, meaning that there are three choices. In the second (and only) position left, any one of the remaining two books can be placed, meaning that there are two choices. Together there are 3 × 2 = 6 possibilities, or permutations. Considering the second case, where there are four books, three of these taken at a time, for the first position, there are four choices, for the second position there are three choices, for the third position (and only one left), there are two choices, together amounting to 4 × 3 × 2 = 24 permutations. We note that this is a product of three factors, consecutive numbers, starting with four.

Along similar lines, the number of possible permutations within a set of n things, taking r things at a time, is

$$n \times (n - 1) \times (n - 2) \ldots (r\text{th factor})$$

The third factor above is $(n - 2)$, the fourth factor will be $(n - 3)$, and the fifth factor will be $(n - 4)$. The rth factor will be

$$n - (r - 1)$$
$$= (n - r + 1)$$

Now, using the designation ${}^nP_r$ for the number of permutations of n things, r of these taken at a time, we may write

$${}^nP_r = n(n—1) \times (n - 2) \ldots \times (n - r + 1) \tag{16.1}$$

If, in the set of n things, all of them are taken at a time $(r = n)$, we have

$${}^nP_n = n \times (n - 1) \times (n - 2) \ldots \times 3 \times 2 \times 1 \tag{16.2}$$

The product of all rational numbers from n to 1 is called the *factorial of n* and is symbolized as $n!$

Then

$${}^nP_n = n! \tag{16.3}$$

Using this symbol, further simplification of (16.1) is possible; multiplying and dividing by the same factor shown in square brackets below, we have

$$^{n}P_{r} = \frac{n\times(n-1)\times(n-2)...\times(n-r+1)[(n-r)\times(n-r-1)\times(n-r-2)...\times3\times2\times1]}{[(n-r)\times(n-r-1)\times(n-r-2)...3\times2\times1]}$$

In this, we note that the denominator is $(n - r)!$, and the numerator is $n!$ Therefore,

$$^{n}P_{r} = \frac{n!}{(n-r)!} \quad \text{(for } r < n\text{)} \tag{16.4}$$

16.8.2 Combinations

Combinations of three things, A, B, C, with two of these taken at a time, are

AB BC CA

The other three

BA CB AC

do not count as combinations, because these are only variations in order. In the second case we considered, with four things, A, B, C, D, three of these taken at a time, the combinations are

ABC BCD CDA DAB

The six variations in each column headed by those, in the group of twenty-four numbers shown earlier, are simply the permutations possible with three things, all of them taken at a time; hence, the number of these is 3! Thus, the number of combinations possible with four things, three of these taken at a time, designated as $^{4}C_{3}$, is related to the corresponding permutation as

$$^{4}P_{3} = {}^{4}C_{3} \times 3!$$

Generalizing along the same lines as above, we may write the number of combinations possible within a set of *n* things, taking *r* things at a time, as

$$^{n}P_r = {}^{n}C_r\, r! \tag{16.5}$$

from which we get

$$^{n}C_r = {}^{n}P_r \div r! \tag{16.6}$$

Further, substituting for $^{n}P_r$ from (16.4), we obtain expression for

$$^{n}C_r = \frac{n\,!}{r\,! \times (n-r)\,!} \quad \text{(for } r < n\text{)} \tag{16.7}$$

16.9 Quantitative Statement of Randomization

To illustrate this point, we deliberately work with a small set. Consider a population of six elements, *a*, *b*, *c*, *d*, *e*, and *f*, out of which we want to have a random sample of two elements.

The possibilities are fifteen subsets:

ab bc cd de ef

ac bd ce df

ad be cf

ae bf

af

This is the same as the number of combinations among six things, taken two at a time, which was previously shown to be $^{n}C_{r}$, in this case,

$$^{6}C_2 = \frac{6!}{2! \times (6-2)!} = 15$$

In any one sampling, any one of these subsets has an equal chance of being picked up. If it is done "at random," each of these fifteen subsets has a probability of 1/15, that is, $1/{}^{n}C_{r}$, of appearing as the sample. In the above list of fifteen, *a* is included in five subsets, which is also true for *b*, *c*, *d*, *e*, and *f*. Each item of the set (or population) has an equal chance of being chosen, which is, indeed, the definition of random sampling. Now, conversely, we may generalize *random sampling* as follows: If r items are chosen from a set (or population) of n items, such that each possible subset has the probability of $1/{}^{n}C_{r}$ for being chosen, the r items so chosen are random samples.

16.10 Sampling Methods

The following are typical, but by no means the only, methods used for sampling. Though there is wide variation among the methods used to suit the circumstances, there are some principles to which all sampling methods should conform to render the sample worthy of statistical analysis. The sample should be *representative* of the population, which is likely to happen if the elements of the sample are collected from the population at random. Only with these conditions, can there be a *probabilistic relation* between an inference drawn from the sample and a corresponding inference likely from the population.

16.10.1 Simple Random Sampling

Let us say there is a set of fifty elements in a population, which comprises all the students in a class, in which there are boys and girls, good and mediocre students, coming from well-to-do and poor families. If fifteen students are to be selected for an experiment using a new method of teaching Spanish proposed to be better than the present method, the method of sampling is fairly simple:

1. Write the name of each student on a sheet of paper, using fifty identical sheets, and fold the sheets in an identical way and put those in a hat.
2. Get an "outsider," say, a student from another class, and ask him or her to pick one folded sheet from the hat.
3. Keep it aside; do not replace it. This one student selected, whoever it is, had no greater chance of being selected than anyone else.

4. Shake well to mix the remaining forty-nine sheets of folded paper; ask the outsider to pick another folded sheet.
5. Keep this sheet with the one already selected. This one who is selected also had only as much chance of being selected as anyone else.
6. Repeat the steps (2) through (4) until the subset—the collection of folded sheets of paper picked as above—has fifteen elements.
7. Open the sheets, and read out the names of those who are in the subset. These are the students who form the sample of fifteen selected from the population of fifty.

The advantage of this method is that no special preparation, except writing the names of fifty students, one on each sheet, and folding those was required; it is direct and quick. The disadvantage arises if, instead of fifty students in a class, five thousand students in the entire school district form the population, and instead of the subset of fifteen, we need to have a subset of five hundred. The work then becomes tedious and time-consuming. Also, this method is suitable only when all the elements of the population are known, that is, when the population is *finite*.

16.10.2 Cluster Sampling

If, for instance, the above-mentioned experiment is to be done with all the ninth graders of the public schools in Massachusetts, the experimenter in this case is quite likely not to know how many ninth graders there are in the state, and in that sense, he or she is dealing with an *infinite* population. Further, the expense of involving the ninth graders of all the schools of the state is prohibitively high, even if the time required can be afforded. In such cases the geographical area of Massachusetts is partitioned (on a map) into smaller *regions*. Numbering each region and using these numbers as population, a subset of smaller number of regions is randomly selected. Each region in the subset is then, in turn, partitioned into several *subregions*, and a subset is randomly selected from these subregions. This process may be continued until a reasonably small number of sub-sub-sub . . . regions is selected. Then, the random selection of students is made from each such sub-sub-sub . . . region, and the collected pool of the students so selected forms the sample.

16.10.3 Stratified Sampling

Simple random sampling and even *cluster sampling* are suitable when the population is homogeneous, like all the students being in ninth grade, when no distinction within that category is called for. Say, instead, the sample required is from the population of all students of a large high school having three thousand students, and the experiment is to find how popular the proposed uniform dress code is compared to the existing no dress code. If the investigator gets to know that there are one thousand freshmen, seven hundred sophomores, seven hundred juniors, and six hundred seniors in the school, and if he decides to have a sample of close to one hundred students, he should plan to select (1,000 ÷ 3,000) × 100 = 33 freshmen. On the same basis, twenty-three sophomores, twenty-three juniors, and twenty seniors should be selected. Further, if he gets to know that the girls' opinion is considerably different from that of boys, he may be required to stratify on the basis of gender as well. If there are three hundred fifty girls and two hundred fifty boys among seniors, he would select (350 ÷ 600) × 20 = 12 girls, thus 8 boys from the senior class. This final phase of the sampling can be done either by simple random sampling or by using a random digits table as described earlier in this chapter. All those so selected, boys and girls, from other grades as well, collectively form the required sample. Evidently, in this method, the population is "stratified" into segments so as to give, consciously, proportionate representation to the obvious variety that exists within the population.

16.10.4 Systematic Sampling

This method is popular among industrial quality control experts, having for their populations discrete products issued from assembly lines or automatic machines. Depending on the production rate, the cost of each product, and the level of stringency of quality assurance, it may be decided, for instance, that for every twenty-five products issued by the assembly line, one product should be picked to form an element of the sample; thus the twenty-fifth, fiftieth, seventy-fifth, and so on, products, collected together, form the subset serving as the sample. For a production schedule of one thousand products in a required period, there will be a sample of forty elements.

16.10.5 Multistage Sampling

Where large-scale sampling is involved, such as the job rating of the U.S. president or predicting the outcome of a national election, it is fairly common to use more than one of the above methods in any sequence at different stages, as is found convenient, in the process of collecting a specific sample; this is referred to as *multistage sampling*.

16.11 Bibliography

Cox, D. R. *Planning of Experiments*. New York: John Wiley and Sons, 1992.

Diamond, W. J. *Practical Experiment Designs*. Belmont, CA: Wadsworth: Lifetime Learning Publications, 1981.

Groninger, L. D. *Beginning Statistics with a Research Context.* New York: Harper and Row Publishers, 1990.

Lee, W. *Experimental Design and Analysis*. San Francisco, CA: W. H. Freeman and Co., 1975.

Wilson, B. E., Jr. *An Introduction to Scientific Research*. Mineola, NY: Dover Publications, 1990.

17

Further Significance of Samples

A peculiar blend of general and particular interests is involved in the pursuit of science; the particular is studied in the hope that it may throw light upon the general.

—Bertrand Russell

More often than not, one may expect the values of a dependent variable in an experiment (or the data from a survey) to form a normal distribution. Suppose they do not; suppose they are too few to be convincing in that respect. The experimenter need not despair. A statistical fact known as the Central Limit Theorem comes to his or her rescue and is the subject of this chapter. The predictive power of the normal distribution curve, referred to in Chapter 15, can further be extended to a very powerful predictive tool known as the *standard normal distribution.* It is illustrated here. Use of a statistical table, "Probability Values of the Standard Normal Distribution," which enables one to answer various kinds of probability questions, is demonstrated with examples. These are essential tools in the hands of an experimenter because he or she has access, almost always, only to samples, not to the populations. This is further explained in Chapter 18.

17.1 Inference from Samples

The very idea of sampling is meaningful only when the population is large. When the population is very large, sampling is done to economize on time and expense. When the population involved is not only very large but also not practically accessible, as, for example, when finding the average height of tenth-grade boys in the United States, sampling is unavoidable. Then, we are left with the only choice of drawing the inferences about the population, based on the statistical measures of samples. In drawing inferences from sampling and applying those to the entire population lies the strength of probability and statistics. When the entire population is accessible and reasonably small, computations for the population of such statistical measures as *mean, variance*, and *standard deviation* are done using the same formulas as in case of samples. But to distinguish statistical measures of samples from those of the population, different sets of symbols, shown in Table 17.1, are very often used in literature.

Table 17.1 *Symbols normally used for statistical measures.*

Sample		Finite Population
n	Size (number of elements)	N
$\bar{x}$	Mean (of the values of the variable)	μ (or $\bar{x}$)
S^2_x	Variance	σ^2
s_x	Standard deviation	σ
p	Probability (of success) of each trial	P

17.2 Theoretical Sampling Distribution of $\bar{x}$

When we encounter finite but large or infinite populations, sampling is inevitable. Suppose we are interested in knowing the *population mean*, the only connection we have with the population being the sample; we have to depend on the *sample mean*, which is not an exact but a "probabilistic" value, meaning it is reasonably close to the population mean. Taking a sample from the population and finding the sample mean is, in effect, like doing an experiment with a given set of conditions. To increase the confidence in the experimental results, it is common to replicate the experiment; the greater the number of replications, the greater the confidence. If the experimental outcome, the dependent variable, is a number, then the number so obtained from several replications will be averaged to get the "final" number as the experimental result. The equivalent in sampling is to get several independent samples at random from the population, to find the mean from each sample, and then to get the mean of these means, a procedure that will increase the confidence (or the likelihood) of the mean of the mean being close to the mean of the population.

For demonstration purposes, we take a small population of only six elements: 4, 8, 12, 16, 20, and 24. If we decide, for instance, to draw four random samples from the population, those will be four out of all the possible combinations. The mean value of the four means, one from each sample, is then the probable mean of the population. By increasing the number of samples, hence, the number of means to be averaged, we expect the probable mean to get even closer. With this in view, we will take all the possible samples, each of four elements, from the population, which is simply all the possible combinations of four elements taken at a time, from a set of six elements. The number of such combinations is given by

$$^{6}C_{4} = \frac{6!}{4! \times (6-4)!} = \frac{6.5.4.3.2.1}{4.3.1.2.1} = 15$$

This means it is possible to draw fifteen different samples. The actual samples and their means are listed respectively in the second and the third columns of Table 17.2. Entries in the fourth column are to be interpreted as follows: When one of these fifteen possible samples is picked at random from the population, the *probability of success* of this selection is 1/15, in the same way that the probability of success of getting a two from a throw of a die is 1/6, because there are six faces, each a possibility, and two is the one that was successful in selection.

Now, if we take a close look at the third and the fourth columns of Table 17.2, we see another relation emerging. The mean value ten, with a probability of 1/15, appears in the list only once among the fifteen possibilities; the same is true for mean values eleven, seventeen, and eighteen. On the other hand, the mean value twelve is found twice, making the probability of its appearance 2 × 1/15 = 2/15; and the same is true for mean values thirteen, fourteen, and fifteen. The mean value fourteen appears three times, making its probability 3 × 1/15 = 3/15. Extracting these facts from Table 17.2, a new relation between the value of the sample mean, $\bar{X}$, and the corresponding probabilities (of suc-

Table 17.2 *Random Samples of a Population (4, 8, 12, 16, 20, and 24)*

Sample#	Elements	Sample mean	Probability (of success) of this sample
1	4, 8, 12, 16	10	1/15
2	4, 8, 12, 20	11	1/15
3	4, 8, 12, 24	12	1/15
4	4, 8, 16, 20	12	1/15
5	4, 8, 16, 24	13	1/15
6	4, 8, 20, 24	14	1/15
7	4, 12, 16, 20	13	1/15
8	4, 8, 16, 24	14	1/15
9	4, 12, 20, 24	15	1/15
10	4, 16, 20, 24	16	1/15
11	8, 12, 16, 20	14	1/15
12	8, 12, 16, 24	15	1/15
13	8, 12, 20, 24	16	1/15
14	8, 16, 20, 24	17	1/15
15	12, 16, 20, 24	18	1/15

cess), designated as a probability function, $f(\overline{X})$, is shown in Table 17.3. Such relations as shown in this table are known as *theoretical sampling distribution* for the population under consideration (finite or infinite).

Table 17.3 *Theoretical Sampling Distribution for* $\overline{X}$

$\overline{X}$	$f(\overline{X})$
10	1/15
11	1/15
12	2/15
13	2/15
14	3/15
15	2/15
16	2/15
17	1/15
18	1/15

This relation is highly significant as we try to understand the nature of the population, with samples serving as the medium. The various statistical properties, both of the central tendency and the distribution, can be worked out reasonably closely through the data provided by theoretical sampling distribution. For instance, there are nine sample means in Table 17.3. The mean of the sample means is, thus, given by

$$\mu_x = (10 + 11 + 12 + 13 + 14 + 15 + 16 + 17 + 18) \div 9 = 14$$

We will separately calculate the mean of the population (which we happen to know entirely). It is given by:

$$\mu_x = (4 + 8 + 12 + 16 + 20 + 24) \div 6 = 14$$

This result is no coincidence. The truth of this "experiment" can be confirmed by considering much larger populations than the one of only six elements considered here. But the gravity of this significance is lost to us because not only do we know all the elements of the population, but also the population is small. If the population, instead, were very large (or infinite), like the height of all the tenth-grade boys in the United States, then to arrive at the *mean height*, we have to depend on sampling alone. Taking several samples and from those preparing a theoretical sampling distribution is the right procedure to follow. But it is to be noted that the particular statistic we so obtain, for instance, the *mean of the population*, can be only probabilistic, not exact as demonstrated above, which was made possible because in that case we knew the entire population.

In an infinite population, the number of samples that can be taken is also infinite; but practically, it is necessarily a finite, and often a limited, number. It can be intuitively understood that the larger the number of samples, and the larger the number of elements in individual samples (these need not be the same from sample to sample), the closer the probable statistic to the corresponding one of the population. Summarized as follows are the procedural steps for preparing the theoretical sampling distribution of sample means when the population is either too large to take all possible samples or it is simply infinite:

1. Take a random sample with a reasonably large number of elements; find the mean of this sample and call it $\overline{X}_1$.
2. As above, take a large number of independent, random samples, find the mean of each sample, and call these $\overline{X}_2, \overline{X}_3, \overline{X}_4, \ldots$.
3. Out of such data, consisting of a conveniently large number of mean values, count the number of times $\overline{X}_1$ appeared and call this $n_{\overline{X}1}$ also, find $n_{\overline{X}2}$, $n_{\overline{X}3}, n_{\overline{X}4}, \ldots$.
4. Find the numerical value of

$$\frac{n_{\overline{x}1}}{\sum n_{\overline{x}1} + n_{\overline{x}2} n_{\overline{x}3} + n_{\overline{x}4} + \ldots}$$

   for all the data.

The above is a fraction expressing how frequently $n_{\overline{X}1}$ appeared among all $n_{\overline{X}2}, n_{\overline{X}3}, n_{\overline{X}4}, \ldots$ put together. It is known as the *relative frequency of* $\overline{X}_1$; we designate it as $f(\overline{X}_1)$.

5. As above, find the relative frequencies of $\overline{X}_2, \overline{X}_3, \overline{X}_4, \ldots$; designate these as $f(\overline{X}_2), f(\overline{X}_3), f(\overline{X}_4), \ldots$
6. Then, prepare a table showing in two columns $\overline{X}_1$ versus $f(\overline{X}_1)$, $\overline{X}_2$ versus $f(\overline{X}_2)$, and so on, for all values of sampling means. This table is the theoretical sampling distribution of sample means for the population under consideration and is similar to Table 17.3.

In passing, we should note that the third column of Table 17.2 shows the mean, $\overline{X}$, of each sample drawn. This is so because that is the particular statistic we chose to calculate for each sample drawn. If we choose to calculate, instead, a different statistic, let us say the *variance*, we should have in that column a value for the variance of each sample. But the probability of each sample, shown in the fourth column of Table 17.2, would not have changed. Extracting data from such a table, we could have constructed a theoretical sampling distribution for variance, as in Table 17.3, but with different numerical values in its two columns, namely, variance, s^2_x, versus probability function of variance, $f(s^2_x)$. In other words, it is possible to construct a theoretical sampling distribution of any statistic we choose, but statisticians prefer and rely more on the theoretical sampling distribution of sample means than on any other statistic.

Using a theoretical sampling distribution of sample means as the source, it is possible to predict, that is, statistically calculate, other statistics of the population besides the population mean. It is to be understood that values so predicted have only probabilistic value, not certainty value. Even so, relations between predicted individual statistics of the population and the corresponding ones worked out using sample means are usually written in the form of equations. Though the empirical "truth" of these can be demonstrated by "experiments" using smaller populations, as was done before for population mean, we will skip that exercise and write the following relations:

$\mu_{\bar{x}} = \mu_x$ for all situations

$$\sigma^2_{\bar{x}} = \frac{\sigma^2_x}{n}\left(\frac{N-n}{N-1}\right)$$

in which n is the number of elements in each of the many samples taken, when N is known

$$\sigma^2_{\bar{x}} = \frac{\sigma^2_x}{N}$$

as above, but when N is infinite or very large

$$\sigma_{\bar{x}} = \frac{\sigma_x}{\sqrt{n}}\sqrt{\frac{N-n}{N-1}}$$

when N is known

$$\sigma_{\bar{x}} = \frac{\sigma_x}{\sqrt{N}}$$

when N is infinite or very large

Revisiting Table 17.3, we need to make the following important observation. In the second column, we have various fractions: 1/15, 2/15, 3/15. The first one, 1/15, and the second one, 2/15, both appear four times, and the third one, 3/15, appears only once. Each of these represents the probability of success of the corresponding sample mean. Adding together all these probabilities, we get 4(1/15) + 4(2/15) + 1(3/15) = 1.0. This is analogous to adding together the probabilities of rolling a one, two, three, four, five, and six, all the possibilities for a given dice throw; as each has a probability 1/6, adding them together yields 1.0. In a practical case of drawing many samples from a very large or infinite population, thereby constructing a theoretical sampling distribution of sample means, likewise we expect the sum of the probability functions to be close to, but not exceed, 1.0. That is,

$$f(\bar{x}_1) + f(\bar{x}_2) + f(\bar{x}_3) + \ldots f(\bar{x}_n) = 1.0$$

17.3 Central Limit Theorem

Now that we have looked at an example of theoretical sampling distribution in action, we are now ready to witness one of the most far-reaching generalizations in the area of statistics. Named the Central Limit Theorem, it may be stated as follows: For a given population, finite, large, or infinite, whatever may be the shape of the frequency distribution, the graphical relation of $\bar{x}$ (the mean value of individual samples from the population, of which several need to be drawn) versus $f(\bar{x})$, the probability function (or relative frequency), approximates to be a normal distribution curve.

It is known that each one of the following factors renders the approximation closer to the ideal:

1. The larger population, N
2. The closer distribution of N to normal
3. The larger number of elements in each sample, n
4. The larger number of samples drawn

This theorem, like the normal distribution curve representing a large population of common significance, is empirical in nature, meaning it may be subjected to "experimentation."

Equipped with the data in Table 17.3, though it is confined to the population of only six elements (4, 8, 12, 16, 20, 24), we are now ready for the experiment, namely, to plot a graph with numerical values of $\overline{x}_1, \overline{x}_2, \overline{x}_3, \ldots$ as abscissa and $f(\overline{x}_1), f(\overline{x}_2), f(\overline{x}_3), \ldots$ as corresponding ordinates. The result is shown in Figure 17.1. Though there are only nine points in the graph, some significant observations and remarks can be made.

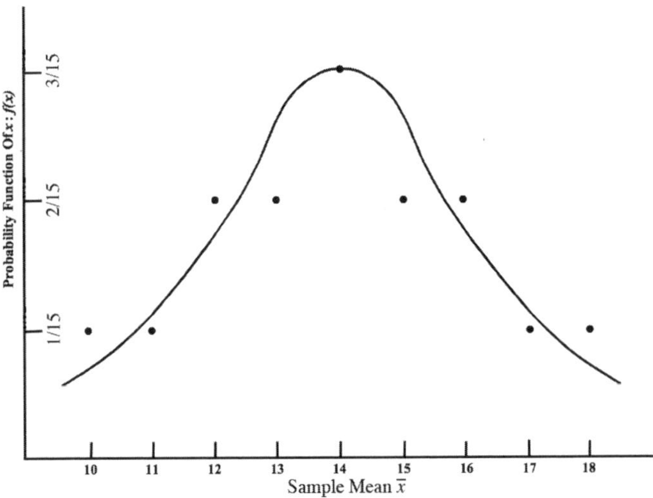

Figure 17.1 *Manifestation of the Central Limit Theorem.*

1. The numerical value of $f(\overline{x})$ is least toward both extremes of $\overline{x}$; it is highest in the midspan of $\overline{x}$. This is one of the characteristics of a normal distribution curve.
2. Led by this fact, we attempt a bell-shaped curve among the nine points. Though the curve does not touch any one of the points, it is fairly close to

each. The attempted curve is, thus, a good approximation to normal distribution. If there were more random variables, that is, more $\bar{x}$ values, made possible by a larger population of N, it would be reasonable to expect a closer fit among the points.

3. Another feature to be noted in this graphical relation is as follows: the ordinates in the normal distribution of a large population show the number of discrete elements, expressed as positive integers, corresponding to the values of variable x. In contrast, we have here ordinates representing probability functions, each a fraction. In the normal distribution of population, all the ordinates corresponding to various "ranges" of x added together give the total number in the entire population, whereas the sum of all the ordinates in the present graph, however many they may be, gives the sum of all the probabilities together, which is 1.0.

The population we started with, namely, 4, 8, 12, 16, 20, and 24, with only one element of each value, if plotted for frequency distribution, yields a flat, horizontal straight line, far from the shape of a normal distribution curve. But the graphical relation $\bar{x}$ versus $f(\bar{x})$, derived from this same population, nonetheless approximates a normal distribution. This is the meaning of the Central Limit Theorem. This theorem, through the means of sampling as a process, serves as a link between the statistics of a given large (or infinite) population and its probability characteristics.

17.4 Standard Normal Distribution

The normal distribution curve, as mentioned in Chapter 15, occupies a preeminent position in statistics. That the frequency distributions of many (though not necessarily all) natural and man-made random variables approximate to this shape is an important reason for its significance, which is further enhanced, as we have seen, by the consequence of the Central Limit Theorem. This significance, in turn, is derived from the mathematical relation to which the normal distribution curve conforms, meaning that the x-y relation of this curve is entirely determined by two statistics: (1) the mean μ_x, and (2) the standard deviation, σ_x, both of which are either known or can be determined from the given set of variables. In other words, a given normal distribution is a function of these two variables, and the shape of that distribution curve is unique; thus, two or more normal distributions having the same means, μ_x, and the same standard deviation, σ_x, are totally identical

17.5 Frequency Distribution and Probability Function

Let us say that statistics were required on the height of adult males in the United States. The usual, established, statistical procedures take over. Because, obviously, it is not practical to reach every adult male in the country, an appropriate random sampling is adapted. The hypothetical data obtained and tabulated in Table 17.4 is where we start.

Table 17.4 *Data on Height of Adult Males; Random Sample*

Height in Inches	No. of Adult Males	Percentage of Total
51.9–55.9	2	0.26
55.9–59.9	9	1.21
59.9–63.9	45	6.03
63.9–67.9	161	21.58
67.9–71.9	286	38.33
71.9–75.9	177	23.72
75.9–79.9	53	7.10
79.9–83.9	7	0.93
83.9–87.9	4	0.54
87.9–91.9	2	0.2
Total	746	

This same data is shown graphically in Figure 17.2, which, as we recall, is a typical frequency histogram. The lines joining the class marks in this figure show the tendency toward normal distribution. If the classes were taken with narrow widths (class ranges), for instance, an increment of 1 inch instead of 4 inches, the tendency toward normal distribution would be even more obvious. We will assume at this point, which is reasonable, that with more individuals in the sample than the present 746 and with narrower increments in height (on the abscissa) in the figure, the data will approximate to normal distribution.

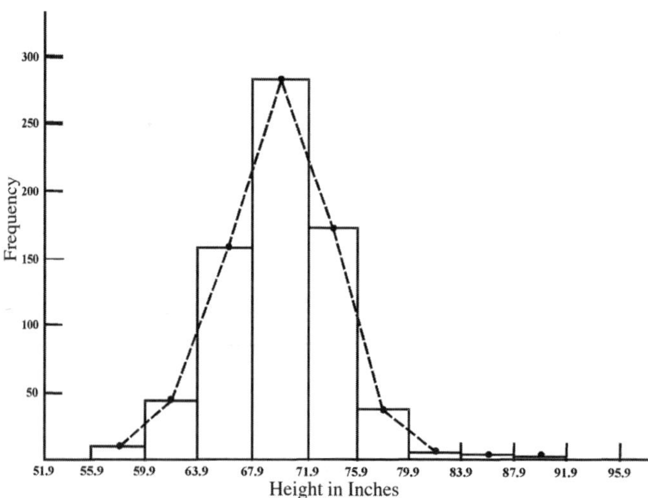

Figure 17.2 *Graphical presentation of data in Table 17.4.*

Thus far, we note, our consideration is confined to statistics, as if it were free of probability. We now recall that in the process of random sampling, the element of chance was operative without hindrance. That there should be a given number in each class, for instance, 161 individuals with height between 63.9 and 67.9 inches, was a matter of pure chance. Stated differently, the probability of finding 161 adult males in the population with the height between 63.9 and 67.9 inches, judging from the present sample, was 161 ÷ 746. On this basis, we can now assign probability values for each class and prepare Figure 17.3 and justifiably call it a probability distribution for the random sample.

The same data are shown graphically in Figure 17.4. The abscissas in this figure, as in Figure 17.2, are classes of height, but the ordinates are not the frequency distribution; instead, they are measures of the probability of finding adult males with the corresponding ranges in heights. Each ordinate in this graphical relation may be considered as the probability function of the corresponding class width; for instance, the probability function of height 71.9 to 75.9 inches is 177 ÷ 746. This probability can be expressed, using p (instead of the conventional f) in the functional relation:

$$p\ (71.9\text{–}75.9\ \text{in}) = 177 \div 746$$

Further, we note that the values of probability functions, given by the height of bars, vary along the classes of the random variable (on the abscissa). This fact may be expressed as the probability having different *densities* as we proceed from one end of the variable to the other. Starting at the lowest density of 0.0026 for the variable at 51.9 to 55.9 inches, the density increases to a maximum of 0.3833 at 67.9 to 71.9 inches, then decreases to the lowest value of 0.0027 at 87.9 to 91.9 inches. Conforming to this observation, Figure 17.4 is more appropriately referred to as a probability density distribution.

For the data in Figure 17.3, we can also plot a *cumulative distribution of probability* curve as shown in Figure 17.4.

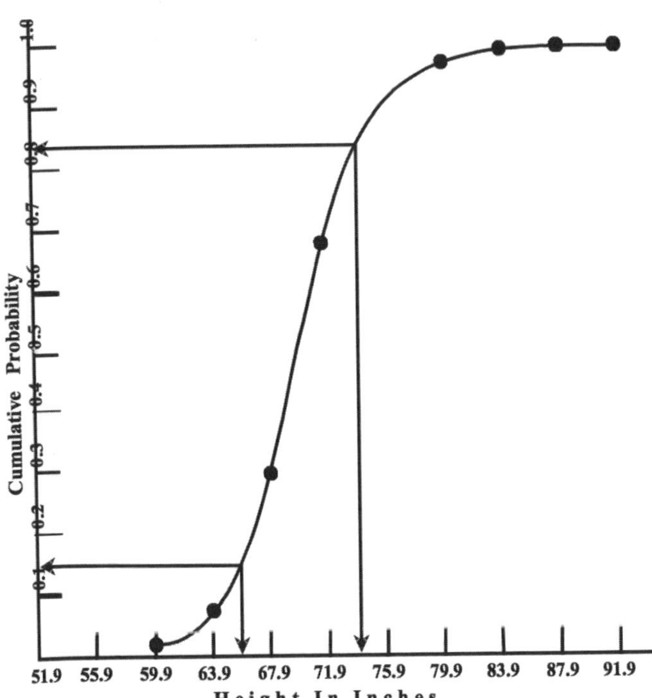

Figure 17.4 *Cumulative distribution of probability.*

Table 17.5 *Probability distribution for the random sample.*

Inches	Probability of success in selection	Cum prob
51.9-55.9	0.0026	0.0
55.9-59.9	0.0121	0.01
59.9-63.9	0.0603	0.075
63.9-67.9	0.2158	0.290
67.9-71.9	0.3833	0.6741
71.9-75.9	0.2372	0.9113
75.9-79.9	0.0710	0.9823
79.9-83.9	0.0093	0.9916
83.9-87.9	0.0054	0.9970
87.9-91.9	0.0027	0.9997

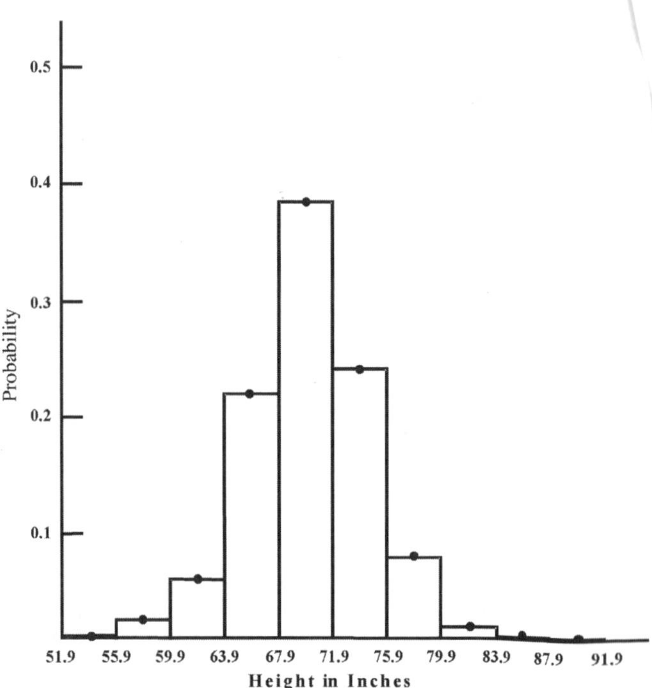

Figure 17.3 *Probability distribution of the random sample.*

This curve is suitable for answering such questions as

1. What is the probability that the height of all adult males in the United States is above 75.9 inches? The answer to this question is 1 – 0.9 = 0.1 (or 10 percent).
2. What is the probability that such height is below 59.9 inches? It is 0.025 (or 2.5 percent).

Note that the sum of all the probabilities (of success) in Table 17.5, which is also the probability corresponding to the maximum height in Figure 17.4, is 1.0. As we pointed out earlier, relative to normal distribution, the number of "bars" in the histogram (Figure 17.3) can be increased by decreasing the class widths (on the abscissa), thereby making the lines joining the probability values tend to be smooth, approximating the shape of normal distribution. Then, the sum of the areas of all the bars, together, which in the limit is the curve, is 1.0. In summary, we may say that the area under the curve of the probability density distribution, or the probability functions of a given random variable in a sample (or in a population), is unity. We also note that the data for the frequency distribution (Figure 17.2) and that of the probability density (Figure 17.4), if both are approximated to normal distribution, present similar curves. The only difference is in the scale of the ordinates. Whereas the ordinates for frequency distribution are in discrete numbers of elements within different classes, for probability density, the ordinates are in fractions or percentages, if we prefer, showing the probabilities of success of different classes.

17.6 Standard Normal Curve

The *standard normal curve* is, in fact, a generalization of many—as many as we may encounter—normal curves showing probability density distributions. Before we move to the generalization involved, it is relevant to examine the nature of the normal distribution curve. Firstly, it is not a frequency polygon, even if hundreds of straight lines are required to connect the points. Similarly, it is not a histogram, even if hundreds of "bars" constitute the figure, but the histogram may approximate the normal

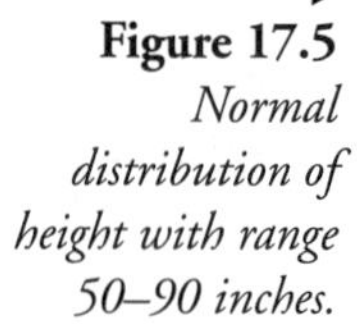

Figure 17.5 *Normal distribution of height with range 50–90 inches.*

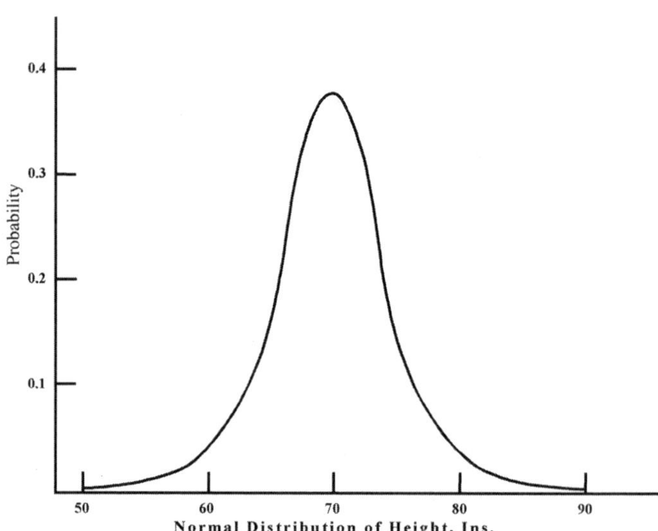

distribution curve, when the class widths are virtually reduced to lines. The normal distribution curve is, indeed, a continuous curve, bell-shaped and symmetrical about the mean of the variable. Though it is obtained purely as the outcome of a statistical procedure, it prevails as an embodiment of quantitative probability. It serves as a means, a tool, by which we can answer many questions of vital importance, such answers being the only ones possible in the absence of certainty. Based on such answers, decisions need to be made in daily business as well as in scientific inference. In way of preparing for the study of the standard normal curve, let us make the following assumptions about data and procedures for obtaining the statistics on the height of adult males in, let us say, England.

1. The range of heights is between 50 and 90 inches, and it was measured to a precision of 0.05 inch. Then, there are (90 – 50) ÷ 0.05 = 800 classes, each of 0.05 inch, for height as the variable.
2. The random sample consists of about ten thousand adult males.
3. A table of probability distribution, similar to Table 17.5, is prepared for the data

4. Using data from this table, a histogram of probability density distribution, similar to Figure 17.3, is prepared.
5. Assuming that the class points at the top end of the bars conformed to a Gaussian distribution, which is very likely, a smooth curve is drawn.

The result then, in graphical form, will be as shown in Figure 17.5.

We note that the area under this curve (which is a measure of the sum of all probabilities) is of numerical value 1.0. Using this figure as the source, we may ask several probability questions.

1. What is the probability that the height of adult males of England will not exceed 63 inches? The answer to that question is given by the numerical value of the area under the curve up to sixty-three on the abscissa.
2. What is the probability that the height of adult males will exceed 75 inches? The answer to this question is given by the numerical value of the area under the curve beyond seventy-five on the abscissa.
3. What is the probability that the height of adult males will vary between 65 and 75 inches? The answer to this question is given by the numerical value of the area under the curve between sixty-five and seventy-five.

Each one of the above three questions, being only part of the area, is a fraction numerically less than one. In contrast, suppose we ask the question, What is the probability that the height of adult males in England will vary only between 50 and 90 inches? The answer is, obviously, the numerical value of the area under the curve between abscissas fifty and ninety. On impulse, we are likely to say it is 1.00, but reflecting on the nature of the normal distribution, we should say that it is very close to but less than

1.00. This is so because, at both ends, the normal distribution curve is asymptotic, extending from less than fifty to more than ninety and never touching the x-axis at either end.

Common to all four answers above is the phrase "numerical value of the area under the curve." Obviously, the next important consideration is how to evaluate the area under the normal distribution curve. Here comes the well-known application of elementary calculus. In the functional relation

$$y = f(x)$$

y takes different values as x varies over a range. If the different values of y, plotted as ordinates corresponding to the different values of x, when connected, form a smooth curve (or line), the area under the curve, corresponding to a range of x values, namely, between x_1 and x_2, is obtained by integrating the function, $f(x)$, between the limits x_1 and x_2, symbolically expressed as

$$\int_{x_1}^{x_2} f(x)dx \qquad (17.1)$$

Now we recall that the normal distribution curve conforms to the functional relation:

$$y = f(x) = \frac{1}{\sigma\sqrt{2\pi}} e^{-1/2(\frac{x-\mu}{\sigma})^2} \qquad (17.2)$$

We have seen that in the graphical form of this relation (Figure 17.5), the area under the curve between any two values of x is a measure of probability. Considering the interval on abscissa from x to $(x + dx)$, we may now write the expression for probability corresponding to width dx as

$$\text{Probability, } p = f(x)\, dx \qquad (17.3)$$

Substituting from (17.2), we get

$$\text{Probability, } p = \frac{1}{\sigma\sqrt{2\pi}} e^{-1/2\left(\frac{x-\mu}{\sigma}\right)^2} dx \tag{17.4}$$

Suppose now that we adapt a new scale for measuring the range on the abscissa, with no other changes in the probability density curve, and this scale is related to x by

$$z = \frac{x - \mu}{\sigma} \tag{17.5}$$

Then,

$$x = \mu + z \, . \, \sigma \tag{17.6}$$

And in terms of *z scale*, the probability is given by

$$\text{Probability, } p = \frac{1}{\sigma\sqrt{2\pi}} \times e^{-\frac{1}{2}(z)^2} \times d(\mu + z\sigma) \tag{17.7}$$

Considering that μ and σ are both constants for a given distribution, we get

$$d(\mu + z\sigma) = 0 + \sigma \, . \, dz \tag{17.8}$$

Substituting $(0 + \sigma \, . \, dz)$ in place of $d(\mu + z\sigma)$ in (17.7), we get

$$\text{Probability, } p = \frac{1}{\sqrt{2\pi}} e^{-\frac{1}{2}z^2} dz \tag{17.9}$$

Comparing (17.9) with (17.4), we notice that (17.9) may be considered a particular case of (17.4), with $\mu = 0$, $\sigma = 1$, and z in place of x. This means that the graphical representation of (17.9) is a probability density curve. Such a curve is known as the *standard normal probability curve* and is shown in Figure 17.7. It is the function of a random variable that has a normal distribution with a mean of 0 and a standard deviation of 1.0. We further note that any distribution of a random variable, so long as it is "normal," can be reduced to this standard shape, represented by (17.9) by making the substitution as in (17.5); z then is known as the *standard normal variable*. And this entails that for a given normal distribution curve, the mean, μ, and the *standard devia-*

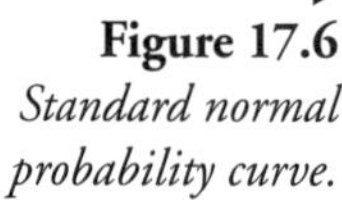

Figure 17.6
Standard normal probability curve.

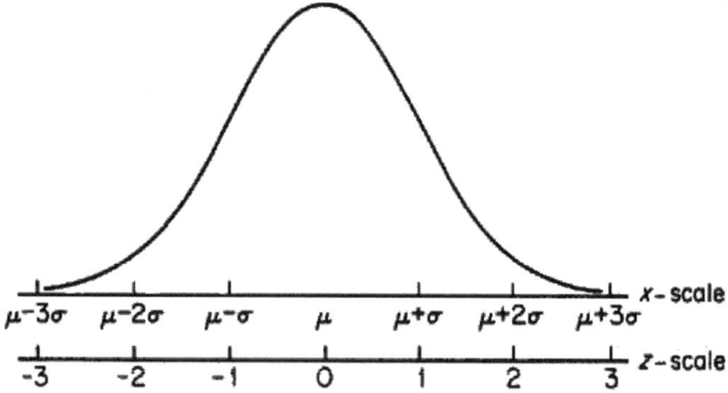

tion, σ, of the random variable should be known before it can be rendered into this "standard" format.

For the purpose of illustrating the use of the standard normal distribution curve, we will ask questions pertaining to two diverse fields of interest:

1. Suppose the marketing department of a screw manufacturing company, on one of its products, a 1-inch screw, accepts the customer demand that screws either shorter than 0.97 inches or longer than 1.04 inches should be rejected. What is the probability that the company can meet this demand immediately without retooling?
2. What is the probability that the newborn babies in the United States have a length of only between 18 and 26 inches?

Question 1 above is from the area of quality control in manufacturing, and question 2 is from the area of health care. Supposing we have a normal distribution curve corresponding to question 1, we can get the answers in terms of the area under the curve, integrating (17.2), between appropriate limits as specified in the question. Depending on the probability density of the random variable, the shape of the curve corresponding to question 2 will be different from that for question 1. To answer any questions relative to this distribution, we need to perform another integration. Add to this that integration of this type of equations is not easy, though considerable help is now available from computers.

Suppose there is a need to integrate just one equation, and the results are applicable to any normal distribution whatsoever. That, precisely, is the equation that represents the standard normal distribution. The integration is already done by the mathematicians; the area under the curve is evaluated for all possible limits to the required degree of accuracy and provided in the form of a compact table, referred to as "Areas of Probability for Standard Normal Distribution" (APSND). It is made available in many standard texts on statistics, as well as in books on other subjects having to deal with probability. It is reproduced here as Table 17.5.

Table 17.6 *Area of Probability for Standard Normal Distribution (APSND)*

z	.00	.01	.02	.03	.04	.05	.06	.07	.08	.09
0.0	.0000	.0040	.0080	.0120	.0160	.0199	.0239	.0279	.0319	.0359
0.1	.0398	.0438	.0478	.0517	.0557	.0596	.0636	.0675	.0714	.0753
0.2	.0793	.0832	.0871	.0910	.0948	.0987	.1026	.1064	.1103	.1141
0.3	.1179	.1217	.1255	.1293	.1331	.1368	.1406	.1443	.1480	.1517
0.4	.1554	.1591	.1628	.1664	.1700	.1736	.1772	.1808	.1844	.1879
0.5	.1915	.1950	.1985	.2019	.2054	.2088	.2123	.2157	.2190	.2224
0.6	.2257	.2291	.2324	.2357	.2389	.2422	.2454	.2486	.2517	.2549
0.7	.2580	.2611	.2642	.2673	.2704	.2734	.2764	.2794	.2823	.2852
0.8	.2881	.2910	.2939	.2967	.2995	.3023	.3051	.3078	.3106	.3133
0.9	.3159	.3186	.3212	.3238	.3264	.3289	.3315	.3340	.3365	.3389
1.0	.3413	.3438	.3461	.3485	.3508	.3531	.3554	.3577	.3599	.3621
1.1	.3643	.3665	.3686	.3708	.3729	.3749	.3770	.3790	.3810	.3830
1.2	.3849	.3869	.3888	.3907	.3925	.3944	.3962	.3980	.3997	.4015
1.3	.4032	.4049	.4066	.4082	.4099	.4115	.4131	.4147	.4162	.4177
1.4	.4192	.4207	.4222	.4236	.4251	.4265	.4279	.4292	.4306	.4319
1.5	.4332	.4345	.4357	.4370	.4382	.4394	.4406	.4418	.4429	.4441
1.6	.4452	.4463	.4474	.4484	.4495	.4505	.4515	.4525	.4535	.4545
1.7	.4554	.4564	.4573	.4582	.4591	.4599	.4608	.4616	.4625	.4633
1.8	.4641	.4649	.4656	.4664	.4671	.4678	.4686	.4693	.4699	.4706
1.9	.4713	.4719	.4726	.4732	.4738	.4744	.4750	.4756	.4761	.4767
2.0	.4772	.4778	.4783	.4788	.4793	.4798	.4803	.4808	.4812	.4817

2.1	.4821	.4826	.4830	.4834	.4838	.4842	.4846	.4850	.4854	.4857
2.2	.4861	.4864	.4868	.4871	.4875	.4878	.4881	.4884	.4887	.4890
2.3	.4893	.4896	.4898	.4901	.4904	.4906	.4909	.4911	.4913	.4916
2.4	.4918	.4920	.4922	.4925	.4927	.4929	.4931	.4932	.4934	.4936
2.5	.4938	.4940	.4941	.4943	.4945	.4946	.4948	.4949	.4951	.4952

Table 17.5 *Area of Probability for Standard Normal Distribution (APSND) (Continued)*

2.6	.4953	.4955	.4956	.4957	.4959	.4960	.4961	.4962	.4963	.4946
2.7	.4965	.4966	.4967	.4968	.4969	.4970	.4971	.4972	.4973	.4974
2.8	.4974	.4975	.4976	.4977	.4977	.4978	.4979	.4979	.4980	.4981
2.9	.4981	.4982	.4982	.4983	.4984	.4984	.4985	.4985	.4986	.4986
3.0	.4987	.4987	.4987	.4988	.4988	.4989	.4989	.4989	.4990	.4990

(Source: Prentice Hall, Inc. 1970, John E. Freund, *Statistics, A First Course*)

We need to develop familiarity with reading this table as a way of answering probability questions. The numerical value of z is read by combining the number in the left-hand column with the decimals of the top line. For example, $z = 1.47$ is obtained by 1.4 in the left-hand column, combined with 0.07 along the top line. Now, looking in the body of the table where the horizontal line of 1.4 meets the vertical column of 0.07, we find the number 0.4292. This is the area of the standard normal distribution between the mean (μ) and $z = 1.47$. It is the corresponding measure of probability, to be stated as, the probability that z varies between 0 and 1.47 is 0.4292. Thus, the area under the curve, between the mean (μ) and any specified positive value of z, can be read; the highest such value is 0.5.

Suppose we want to get probability corresponding to negative values of z. All we need to remember is that the normal distribution curve is symmetrical, the shape of the curve to the left of the mean being a mirror image of the curve to the right of it. Conforming to this, the area for the value $z = -2.38$, for instance, is 0.4913. And the probability value between $z = -1.26$ and $z = +1.26$ is given by $(0.3962 + 0.3962) = 0.7924$.

If we need to find the probability that z is higher than 2.45, we find the corresponding number in the table to be 0.4929.

Since 0.5 is the total area for z higher than the mean (μ), the area for z higher than 2.45 is (0.5 – 0.4929) = 0.0071; this is the measure of the corresponding probability.

Finally, suppose we need to find the probability that z is between 1.25 and 2.5. The area between z = 0 and z = 2.5 is 0.4938, and the area between z = 0 and z = 1.25 is 0.3944. Therefore, the area, the measure of probability required, is (0.4958 – 0.3944) = 0.1014.

17.7 Questions/Answers Using the APSND Table

Now that we are acquainted with the numbers in Table 17.5, we may try to answer the questions we posed before and related ones. A standard normal curve implies normal distribution. This means the data we have on hand should conform, in terms of frequency distribution, reasonably closely to normal distribution. If the population is very large or infinite, a sample of suitable size is collected using an appropriate method (or combination of methods), as described in Chapter 16. Elements in the sample set, in such a case, should conform reasonably to normal distribution. To be able to read the probability values in terms of APSND as a way of answering questions, we need to have the numerical values of just two statistics: the mean (μ) and the standard deviation (σ) of the random variable. Computation of these is fairly easy, as even hand-held calculators are capable of this. Once the computation is done, it is not necessary to have a plot of the data for answering the various questions. Here lies the "ready reference" nature of the standard normal curve, with the associated APSND table.

For purposes of illustration, we assume that the data we have on hand are of original population and, further, that the mean and the standard deviation of the random variables are known.

Example 1:

(Question 1 stated in Section 17.6)

Data:

Mean (nominal length): 1.0 in.

Standard deviation, assumed to be known:0.22

$x_1 = 0.97$ $x_2 = 1.04$

1. $z_1 = \frac{0.97 - 1.0}{0.22} = -1.36$ $z_2 = \frac{1.04 - 1.0}{0.22} = 1.81$

2. For $z_1 = -1.36$, read from table, APSND: 0.4131

   For $z_2 = 1.81$, read from table, APSND: 0.4649

3. Area corresponding to the question: total area on both sides of $z = 0$, that is, $(0.4131 + 0.4649) = 0.8780$

   Answer: The required probability is 87.8 percent.

Example 2:

(Question 2 stated in Section 17.6)

The length of full-term newborn babies in United States is a matter of national statistics, but we will assume the following data:

	Inches:
Minimum	12
Maximum	30
Mean	20
Standard deviation	3.5

Following are the steps for the answer:

1. $z_1 = \frac{18 - 20}{3.5} = -0.57$ $z_2 = \frac{26 - 20}{3.5} = 1.71$

2. For $z_1 = -0.57$, APSND: 0.2157

   For $z_2 = 1.71$, APSND: 0.4564

3. Interpreting the above numbers, the area corresponding to the question: the total area on both sides of $z = 0$, that is, $(0.2157 + 0.4564) = 0.672$ (approx.)

4. Answer: The required probability is 67.2 percent.

We may close this chapter with a reminder that to be able to use this powerful tool, the standard normal probability function, we do not need to have a plotted normal distribution curve. To be able to answer the relevant questions, we only need to know the range of the variable and the corresponding numerical values of μ and σ. However, unless these values have been derived from a population of a reasonably large number of elements or from a reasonably large sample when the population is very large, we cannot develop confidence in the answers. Besides, it is imperative in the function $y = f(x)$ that the values of y are such that the lines connecting these offer a reasonable approximation to normal distribution. This needs to be confirmed. Any and every $y = f(x)$ relation will not necessarily offer a tendency to normal distribution (see Chapter 15).

17.8 Bibliography

Anderson, D. R., D. J, Sweeney, and T. A. Williams. *Introduction to Statistics: An Applications Approach.* St. Paul, MN: West Publishing Co., 1981.

Fischer, F. E. *Fundamental Statistical Concepts.* New York: Harper and Row Publishers, 1973.

Wadsworth, G. P., and J. G. Bryan. *Introduction to Probability and Random Variables.* New York: McGraw-Hill Book Co., 1960.

18

Planning the Experiments in Statistical Terms

Statistical control opened the way to engineering innovation. Without statistical control, the process was in unstable chaos, the noise of which would mask the effect of any attempt to bring improvement.

—*W. Edwards Deming*

For a given project of experimental research, the setup, the instrumentation, and the materials need to be carefully planned; that is obvious. This chapter deals with planning of a different kind: planning the logical basis for the experiment, which is not visible, hence, not so obvious. The logic calls for making statements involving the change in the dependent variable, which is the focus of most experimental research. The statements need to be made in such a way—even before beginning the experiment—that the experimental outcome serves as a means of deciding the truth-value of the statements. But then, the true–not true decisions are to be made in terms of probability, not certainty, because, as we saw in Chapter 15, experimental observations, by their very nature, cannot be proved in any other way than by doing more experiments. In statistical methods, the statements are known as null and alternate hypotheses. Starting with an explanation of these, this chapter details some of the basic, well-known procedures for planning for experiments in statistical terms.

18.1 Guiding Principles

Before building, whether a bridge or a machine, planning is necessary. Likewise, before conducting it, an experiment needs to be planned. A well-planned experiment should be the shortest means (both in terms of time and money) to the ends, answering specific questions asked in the form of hypotheses. Planning for an experiment with quantitative parameters requires an analysis in terms of probability and statistics. This is occasioned by the fact that experimental findings, in general, and measurements, in particular, are not certainties: they are probabilistic in nature. Designing a highly sophisticated experiment requires the help of trained statisticians. This chapter is meant to present the basic outlines involved in the design for students and experimenters who are not specially trained in statistics. The methods of analysis dealt with here are elementary. Nonetheless, they are adequate for planning the simple, comparative experiments required of undergraduate and graduate students, as well as most investigators in industrial research and development. The essential characteristics of designed experiments follow:

1. An experiment should be considered a project, and like an engineered project, it should have well-defined goals. For instance, the phrase "to accomplish as much as possible" is contrary to planning. The quantum of accomplishment should be clearly stated.
2. The planning done at the beginning, middle, and end of the project should have different objectives. In the beginning, it should aim to identify all the possible variables, in the middle, to identify all the significant ones, and at the end, to test for the effect of the most significant ones.
3. The range of each variable should be carefully considered. In the initial stage, testing the effect of an independent variable at more than one level may be unnecessary; in the middle stages, testing at two levels may be adequate; only at the final level can multilevel experiments (with a limited number of variables) be justified.
4. When several independent variables are found significant, some variables may "interact" synergistically with others to produce either "good" or

"bad" effects on the outcome, the dependent variables. Such effects should be identified in the early stages, and planning should be done for specific experimentation in the final stages. The experimenter should specify how much information is adequate at different stages; to say "all that can be known" is contrary to the idea of planning.

5. Each item of information should be assigned a specific level of confidence with the understanding that to expect a 100 percent confidence is to ask for the (statistically) impossible.

6. The experimenter should anticipate the kind of data to be harvested as a result of the experiment. These data should be considered as a means to answering the questions posed by the hypothesis.

18.2 Some Preliminaries for Planned Experiments

The discussion in this chapter is limited to the so-called simple comparative experiments. The purpose of such experiments is to bring about "improvements" in the dependent variable (measured as the experimental response) by causing the planned changes in the independent variables. A comparison is made between the two means (of the numerical values), one before and one after the experiment. Whether the intended improvement did or did not take place is decided based on such a comparison. Also, such a comparison is often used for selecting one of the available two populations with a certain parameter as the criterion, common to both the populations, measured and expressed in numbers.

18.2.1 Sample Size

When the populations are very large or unlimited, sampling is necessary. On the other hand, when the population is limited or reasonably small, sampling becomes less necessary. In planned experiments, every replication of the experiment with the altered (or adjusted) independent variables provides one value of the (hopefully) improved dependent variable. The collection of a number of such values constitutes the "size" of the sample. The larger the sample size demanded, the more replications required. Too many replications result in unnecessary expense, and too few are inadequate for getting a dependable value of the sample mean. The right sample size, thus, needs to be found from statistical considerations.

However, exceptional situations exist. For example, in a hypothetical case, the experimenter, altering the independent variable serving as the "cause," can bring about an increase in the pressure, the "effect," in a steam chamber. Once the independent variable is adjusted as planned, pressure in the steam chamber simply needs to be recorded over an extended period of time (to neutralize the effect of uncontrolled variables, also known as *noise*.) Then, with one replication, a large number of pressure readings can be obtained; these serve as the population. Randomized sampling from such a population needs to be done strictly following statistical considerations. But this does not involve the expense of a large number of replications.

Computation of the sample size, either as the number of replications or as the number of elements randomly selected from a population, involves the following factors:

1. α and β risks
2. δ, the increment of improving a given property as the dependent variable
3. σ, the standard deviation, either known or unknown
4. The nature (or statement) of the null hypothesis for the experiment

Formulas for computing the sample size to suit different experimental situations vary. The experimenter needs to use the formula prescribed by statisticians for a specific set of experimental situations.

18.2.2 Minimum Acceptable Improvement

In a planned experiment, the degree of improvement measured as the enhanced value of the dependent variable needs to be stated.

For instance, in a planned experiment for upgrading the life of a given ball bearing in a specific application, the goal should not be "to improve the life of the bearing to the maximum extent possible." Instead, it is desirable, for example, to state, "to improve the life from the present average of ten thousand hours to the desired average of eleven thousand hours of satisfactory life." All these ideas are implied in what are known as the *hypotheses*, which are stated in symbolic form, containing none of the above words.

18.3 Null and Alternate Hypotheses

Asking the questions and forecasting the likely answers from the experiment are done in the formal language of statistics. The question-answer formats are known as null and alternate hypotheses. These are described below.

18.3.1 Null Hypothesis in an Experiment

The term *null* connotes negation, that is, rejecting (or saying no to) a specific assertion. In the context of experimental research, the *null hypothesis* is used when the experimental data (which is represented by a sample) does not necessarily warrant a generalization (which represents the entire population) that an intended improvement in the dependent variable did not occur. The null hypothesis, H_0, is stated as

$$H_0 : \mu_1 = \mu_0,$$

which is the symbolic way of saying that the independent variable, targeted to be improved, remains unchanged. The failure to improve may be the result of an unaccountable experimental (including sampling) error due to random influences. The experimenter then resorts to a decision-making criterion; for instance, if the probability of the effect of random influences is found to be less than 5 percent, he will reject the hypothesis that the data does not warrant generalization, meaning that the data then is worthy of generalization. The situation may be understood as follows: The experimenter starts with the "pessimistic approach" of saying that the specific data (in the form of a bunch of numbers of common significance) is not fully trustworthy and, hence, deserves to be scrapped. This is accepting the null hypothesis. Then, after careful deliberation in terms of statistical criteria, he may find, for instance, that there are fewer than one in twenty (5 percent) chances that random influences are responsible for what appears to be a meaningful variation that can be deciphered from the data. And so, in moderation, he will develop confidence in the data and make a decision that the data, after all, is not untrustworthy. This is rejecting the null hypothesis.

To the experimenter, the data is all-important. That is what he planned for and strove to get. Once the data is in hand, however, he does not accept it without subjecting it to the "acid test" of statistics, which consists of first taking the side of the critics, and then, only after examining the evidence, being convinced that there is overwhelming reason to move away from the critics and to take, instead, the side of the supporters. This is the most common and convenient path available to the experimenter as a way of developing confidence in his data. The experimenter, thus, states his null hypothesis in such a way that the rejection of the null hypothesis is what he hopes for.

There is in this path a parallel to the treatment of a crime suspect in the court of law. The starting point for the court is to assume that the person accused of the crime is innocent. The prosecutor is required to prove by appropriate evidence that this assumption is wrong. If the evidence is overwhelming against the assumption, the court will change the opinion it started with, that is, reject its own assumption. If, on the other hand, the evidence provided by the prosecutor is not overwhelming, the court will accept its own assumption as valid.

What we have referred to so far as "overwhelming evidence" is not the evidence that may be considered, for instance, as "105 percent strong." There is no such thing in the domain of probability and statistics; there is nothing even "100 percent strong," for then, we would be talking in terms of absolute certainty, which is contrary to statistical truth. The highest possible "strength" in probability approaches, but never reaches, 100 percent. Likewise, there is nothing like 0 percent strength either; the lowest strength approaches, but never reaches, 0 percent. In light of this, within the domain of probability, the data that lead toward accepting the null hypothesis are not to be considered totally worthless. The possibility is open to consider whether there is an alternate method by which the data can be subjected to scrutiny afresh. Appropriately named, there is such a thing as an alternate hypothesis.

But first let us look at and contrast the consequences of either rejecting or accepting the null hypothesis. The parallel that we have drawn between the null hypothesis in logic and the way the court of law deals with the crime suspect can be extended further. Despite all the care exercised by the legal system, it is not impossible that there will be cases in which innocent persons are declared guilty and punished and cases in which guilty persons are declared innocent and set free. In terms of logic, when the null hypothesis is rejected, even though it is true, it is similar to punishing an innocent person; it is called in logic a *type I error*. When the null hypothesis is accepted, even though in reality the hypothesis is false, it is called a *type II error*, which is similar to a guilty person going scot-free. Are the legal system and inferential logic, then, built on faulty foundations? Not quite. The possibility that such errors can occur should be attributed to the fact that the legal system and inferential logic are both based on probability, not certainty; the numerical value of probability can never be 1.0, for then certainty would have asserted itself in the place of probability.

18.3.2 Alternate Hypothesis

The phrase *alternate hypothesis* is deliberately used, with this significance: that if the null hypothesis is proved false, the alternate hypothesis is accepted automatically to be true without the need for a separate and independent proof. The two hypotheses, null and alternate, should be so stated that they conform to this understanding. Thus, the most general way the alternate hypothesis can be stated is

$$Ha : \mu_1 \neq \mu_0$$

With the null hypothesis formulated as $\mu_1 = \mu_0$, the value of μ_1 is definite, since the experimenter knows the value of μ_0. In contrast, with the alternate hypothesis formulated using the sign of inequality ("≠"), the value of μ_1 is indefinite, meaning it may be either less than μ_0 or more than μ_0. Such hypotheses are aptly termed *two-sided alternate hypotheses*. Stating the hypotheses, instead, in the two variations

1. $H_a^1: \mu_1 > \mu_0$
2. $H_a^2: \mu_1 < \mu_0$

(in both cases the reference being μ_0, and the focus of change being on μ_1) will render them "one-sided," though they will remain indefinite, meaning, in either case, the difference between μ_1 and μ_0 is not specified. And, in both the variations, μ_1 can be an improvement over μ_0. It may appear surprising that $\mu_1 < \mu_0$ is an alternate hypothesis, implying improvement, while the numerical value of the property is reducing. The explanation is simple: if the property under consideration is the life of ball bearings in hours, (1) is an improvement, hence the intention of the experimenter. If, instead, the property in question is the percentage of items rejected in the process of performing quality control for a manufactured product, then (2) is an improvement, hence, again, the goal of the experimenter.

As pointed out earlier, however, it is not enough to say there should be "some improvement" in the property. Planning the experiment requires that the degree of improvement should be specified. Then (1) and (2) above should be modified as

3. $H_a: \mu_1 = \mu_0 + \delta$
4. $H_a: \mu_1 = \mu_0 - \delta$

as the case may be.

18.3.3 Risks Involved: α and β Errors

Occasioned by the facts that (1) the null hypothesis will be accepted or rejected on the basis of the sample mean, not the population mean, and (2) the sample mean cannot be expected to be exactly equal to the population mean, it is reasonable not to expect absolute certainty in deriving the inference from the experimental results; there is always an element of risk involved. Relative to the null and alternate hypotheses, the risks are distinguished as follows:

1. The experimenter accepts the alternate hypothesis (H_a) as true, that is, rejecting the null hypothesis, when, actually, the null hypothesis (H_0) is true. In statistics, this is known as an *alpha error*, often simply indicated as α.
2. The experimenter accepts the null hypothesis (H_0) as true, when, actually, the alternate hypothesis (H_a) is true (and the null hypothesis is false). In statistics, this is known as a *beta error*, indicated as β.

We note that an experimenter cannot make both the errors together, for, as mentioned above, it is always either-or between the null and the alternate hypotheses. The risk of making either of the two errors is expressed in terms of probability. For instance, if it is assumed in the design that the experimenter is likely to make an α error of 0.1, then there is a 0.1 (100) = 10 percent chance of his accepting the alternate hypothesis as true, although, actually, the null hypothesis is true, and the chance of his making the correct decision of accepting the null hypothesis is reduced from 100 percent to (1 – 0.1) × 100 = 90 percent. Whether the probability so chosen should be 0.1, as mentioned above, or some other number (such as 0.16 or 0.09) is more or less a subjective decision, made by the experimenter, taking into consideration such factors as (1) the difference it makes, in terms of time and money involved, (2) the level of accuracy intended in the experiment, and (3) how critical the outcomes of the decision are, as applied to a real-life situation.

It is also possible that the experimenter will make the right decision in accepting either the null hypothesis or the alternate hypothesis, as the case may be. But here, again, that he may do so is not an absolute certainty; it is only a probability, however high it may be. Thus, we may think of there being four possibilities with the single-sided alternate hypothesis, each assigned the corresponding probabilities. Table 18.1 shows these possibilities, with the following, normally accepted meanings for the symbols:

μ: the property that is meant to be improved by making certain changes in the related independent variables; understood to be the mean of a set of readings or measurements or calculated values, expressed as a number

δ: the number expressing the quantity or degree of improvement expected in μ

α: *α error*; a probability value (a number) assigned in the process of "designing" the experiment

β: *β error*; a probability value (a number) assigned in the process of designing the experiment

Table 18.1 ***Decision Making; α and β Errors***

The eventual truth to be found by experiment, but unknown at the time of planning	The possibility that the experimenter may decide that there is …	
	No improvement in μ	Improvement in μ
No improvement in μ	Correct decision; probability of this decision: *$(1 - \alpha)$*	Wrong decision; *α error* probability of this decision: α
There is improvement in μ; quantity of improvement: δ	Wrong decision: *β error* Probability of this decision: β	Correct decision; probability of this decision: *$(1 - \beta)$*

Referring to the example on ball bearings in Section 18.2.2, the value of δ, namely, one thousand hours of additional service life, should be specified as $\mu_1 = \mu_0 + \delta$. Its significance lies in that, besides being a criterion to distinguish the null hypothesis from the alternate hypothesis, it relates to the *sample size*, discussed below, which, indeed, is another important factor to decide on in the process of planning. We note that formulating the null hypothesis is fairly easy: it is a statement to reflect that the effort to improve upon the given property is going to be fruitless. But the alternate hypothesis, as pointed out earlier, can be any one of the following statements:

H_a: $\mu_1 > \mu_0$

H_a: $\mu_1 < \mu_0$

H_a: $\mu_1 \partial \mu_0$

In these statements, μ_0 and μ_1 are the mean values of the property in question, meant to be obtained from the entire populations, μ_0 from production before, and μ_1 from the population after, the improvement. But in most (if not all) experiments, only the samples can be subjected to experimentation, not the entire populations. For instance, in the experiment on ball bearings, if it is in the plan to subject every one of the bearings made to the test in lab for its service life, there will be no bearings to sell. It is in this connection that the sample, and through that, the sample mean, $\overline{X}$ (to be distinguished from the population mean), of the property and the sample size enter into the design considerations.

18.3.4 Sample Mean $\overline{X}$: Its Role in the Design

The sample mean, $\overline{X}$ (yet to be obtained and possible only after experimentation), and the population mean, μ_0 (already known prior to experimenting), are logically related as follows:

1. If $\overline{X} \leq \mu_0$, the property in question did not improve; hence, accept the null hypothesis.
2. If $\overline{X} \leq \mu_0 + \delta$, the property did improve per target; hence, accept the alternate hypothesis.

In either case, the decision is simple. But what if $\overline{X}$ is between μ_0 and $\mu_0 + \delta$? That is,

$$\mu_0 < \overline{X} < \mu_0 + \delta$$

Depending on the closeness of $\overline{x}$ either to μ_0 or to $\mu_0 + \delta$, the situation could prompt the experimenter to accept either the null hypothesis or the alternate hypothesis.

18.3.5 Hypotheses Based on Other Parameters

We need to note that, depending on what the experiment sets out to "prove," null and alternate hypotheses may often be stated in terms of other statistical parameters, such as standard deviation or variance, as below:

Null hypothesis H_0: $\sigma_1^2 = \sigma_0^2$ or $\sigma_1 = \sigma_2$

Alternate hypothesis H_a: $\sigma_1^2 \neq \sigma_0^2$ or $\sigma_1 \neq \sigma_0$

But in all the discussions that follow in this chapter, μ, the population mean, is used as the statistic of reference.

18.4 Accepting (or Rejecting) Hypotheses: Objective Criteria

At this stage, the experiment has yet to be conducted and the experimenter has yet to decide which of the two hypotheses he should favor. But the "favoring" is not a matter of fancy; it cannot be decided arbitrarily. The consequences of accepting one or the other of the possible hypotheses should be logically analyzed through the medium of statistical quantities. An outline of such an analysis follows.

Firstly, $\mu_1 = \mu_0$ means that the property in question, as obtained by changing one or more parameters, remains unaltered. This is the null hypothesis, and the two forms of the alternate hypothesis are

1. $\mu_1 = \mu_0 + \delta$
2. $\mu_1 = \mu_0 - \delta$

For one thing, in either form, we cannot say that the improvement will be acknowledged only if it is exactly δ. A little less or a little more should be acceptable. But how much is "a little"? Suppose the intended improvement in the design is 10 percent. Then, should

$$\mu_1 = \mu_0 + 9.9\ (\mu_0)\ \text{percent}$$

or

$$\mu_1 = \mu_0 - 9.9\ (\mu_0)\ \text{percent}$$

whichever is applicable, be rejected as no improvement, thus, treated like the null hypothesis? Doing so amounts to committing a *β error.* How can this error be avoided despite the numbers tempting us to make it? Bound as the experimenter is to work in the domain of probability, not certainty, we may say that he should increase the probability, as much as possible under the circumstances, of not making the error. It is in this form that probability enters the scene. Furthermore (this is important), in the above relations we have used μ_1, the population mean of the improved property, as if it would be accessible. If we could directly measure it, we could compare it with the value $\mu_0 + \delta$ or $\mu_0 - \delta$, as the case may be, see how close μ_1 is to the designed values, and make the appropriate decision. But what would become accessible, after performing the experiment, is only the sample mean, $\overline{x}_1$, which, at best, is only an approximation to μ_1. Now the question is reduced to, How will $\overline{x}_1$, obtained after the improvement, be related to μ_0, which is available before the experiment? Further, $\overline{x}_1$ being the mean for a definite number of elements in the sample, what should be the size of the sample? This becomes another question in designing the experiment. The logic and statistics involved in answering such questions are quite complex and beyond the scope of this book. In this context, the experimenter, assumed to be not a trained statistician, would do well to take on faith, without insisting on proofs, the procedural steps, including do's and don'ts, formulated by statisticians. Four typical designs, each with a distinct set of experimental situations frequently encountered, are discussed as follows.

18.5 Procedures for Planning the Experiments

With minor variations, the following are the steps for designing simple comparative experiments:

1. Identify whether σ is known for the responses represented in μ_0. If it is not known, more calculations will be required, as we will see later in the procedure.
2. State the null and the alternate hypotheses to represent the experimental conditions. Identify whether the alternate hypothesis is "one sided" or "two sided."
3. Make a decision on the appropriate *risk factors*, α and β.
4. Go to Table 18.2 if "one sided" or to Table 18.3 if "two sided"; these tables list the "probability points" appropriate to the situations in the experiment.

Table 18.2 *Probability Points of the Normal Distribution; One-Sided; σ Known*

ρ	
(α or β)	*U*
0.001	3.090
0.005	2.576
0.010	2.326
0.015	2.170
0.020	2.054
0.025	1.960
0.050	1.645
0.100	1.282
0.150	1.036
0.200	0.842

Table 18.2 *Probability Points of the Normal Distribution; One-Sided; σ Known (continued)*

0.300	0.524
0.400	0.253
0.500	0.000
0.600	-0.253

(Source: Wadsworth, Inc., 1981,William J.Diamond, *Practical Experiment Designs for Engineers and Scientists*)

Table 18.3 *Probability Points of the Normal Distribution: Two-Sided; σ Known*

ρ	
(α only)	U
0.001	3.291
0.005	2.807
0.010	2.576
0.015	2.432
0.020	2.326
0.025	2.241
0.050	1.960
0.100	1.645
0.150	1.440
0.200	1.282
0.300	1.036
0.400	0.842
0.500	0.675
0.600	0.524

Corresponding to the numerical values selected for α and β above, find the *probability points*, symbolized here as U_α and U_β.

5. Compute *N*, the *size of the sample*, to be tested for evaluating μ_1.

6. Compute *X*, the *criterion numbers.*

For computing both *N* and *X*, formulas are prescribed appropriate to the set of experimental conditions.

....

Steps that come after this point depend on the "responses" from the experiment attempting to effect the intended improvement, represented by μ_1; but the procedure for using these responses, which are yet to come, can and should be planned in advance.

....

7. Experiment to obtain a sample of *N* elements.

8. Calculate the mean, $\bar{X}_1$, of the improved responses, represented by μ_1, using the prescribed formula, appropriate to the experimental conditions.

9. Compare the numerical values of $\bar{X}_1$ with that of *X.*

10. Decide to accept or reject one or the other of the hypotheses, as prescribed, again, depending on the experimental conditions.

....

The above procedure for planning the experiment is used as the "template" in which each of the steps, 1 to 10, represents a particular function. For example, step 5 is assigned to calculating *N*, the sample size. When a particular step, for instance 5, in turn requires two or more substeps, 5a, 5b, etc. are assigned to such substeps. And, if a particular step, for instance 4, requires two or more iterations, they are shown with subscripts, $(4)_1$, $(4)_2$, etc. Unless otherwise mentioned, the numbers confirm to this scheme in the following discussion including examples.

18.5.1 Criterion Values

For solving problems in designing experiments, statisticians have provided the formulas needed to compute the numerical values of what are known as *criterion values*. The factors involved in the formulas are

σ the standard deviation of one or both populations, as required

N the sample size

δ the minimum acceptable improvement as used in the alternate hypothesis

U_α, U_β the *probability points* (also known as the *standard deviates*) to be found in the appropriate statistical tables

Formulas for computing the criterion values for different sets of experimental situations vary. For a specific set of situations, after computing, the criterion value is kept ready to be compared with the sample mean of the improved responses (the dependent variable), to be obtained after the experimental runs are performed.

Situation Set I

a. It is a test for simple comparison, before and after the improvement.

b. The alternate hypothesis is H_a: $\mu_1 > \mu_0$.

c. μ_0 for population before improvement is known.

d. σ, the standard deviation of the measurements used for μ_0, is known.

With this set of situations, which is more often encountered than others, an example follows:

Example 1:

The following experiment is designed to allay the concerns of an engineer who is required to supervise the performance of a heating plant in a public building. The independent variable available to him is the operating temperature, which can be set as required by a thermostat. The dependent variable is the pressure in a steam vessel, an important component of the heating plant. With the operating temperature set at 300°F, the average pressure reading is 250 psi; the standard deviation is known to be ten. The engineer is facing the demand from his crew to increase the pressure to 260 psi. Based on his experience with the plant and its operation, the engineer predicts that setting the temperature at 350°F will accomplish this goal.

Procedure:

1. σ, the standard deviation, is known to be ten.

   Note: Dimensional units associated with the numbers are suppressed to conform to the format (and brevity) of the hypotheses.

2. State the hypotheses:

   Null hypothesis, H_0: $\mu_1 = \mu_0$

   Change in the pressure planned: 10

   Alternate hypothesis, H_a: $\mu_1 = \mu_0 + 10$

   This is a one-sided hypothesis.

3. Decide the risk factors.

   The risks involved are considered normal. Values, very often used in such analysis, are chosen.

   $\alpha = 0.05$, $\beta = 0.10$

4. Go to Table 18.2.

Corresponding to $\alpha = 0.05$, we find $U_\alpha = 1.645$.

Corresponding to $\beta = 0.10$, we find $U_\beta = 1.282$.

5. Compute N, the sample size, using the formula prescribed for this set of conditions:

$$N = (U_\alpha + U_\beta)^2 \times \frac{\sigma^2}{\delta^2} = (1.645 + 1.282)^2 \times (10^2 \div 10^2) = 8.57$$

Take it as $N = 9$

6. Compute X, the criterion number:

$$X = \mu_0 + \frac{\sigma \times U_\alpha}{\sqrt{N}} = 250 + (10 \times 1.645) \div \sqrt{9} = 255.5$$

....

X is kept ready to be compared with $\overline{x}_1$ (yet to be evaluated after running the experiment), using nine samples.

The remaining part of the procedure is "road testing" the hypothesis.

....

7. Conduct the experiment, which consists of setting the temperature to 350°F, followed (after sufficient time to stabilize the system) by recording several pressure readings over a reasonable length of time.

8. Compute $\overline{x}_1$, the mean of nine randomly selected pressure readings from the population of such readings.

9. Compare $\overline{x}_1$ with X.

10. Make decisions as recommended:

    If $\overline{x}_1 > X$, accept the alternate hypothesis,

    $\mu_1 = \mu_0 + 10.$

    This means, put the plan into action; namely, set the operating temperature at 350°F, which "very likely" will result in the pressure of 260 psi, at the level it is required. "Very likely" here means a probability of 1 – 0.05, that is, 95 percent.

    If, instead, $\overline{x}_1 < X$, accept the null hypothesis,

    H_0: $\mu_1 = \mu_0$, meaning that changing the operating temperature, likely, will not make any change in the pressure. "Likely" here means a probability of 1 – 0.1, that is, 90 percent.

Example 2:

Suppose the null hypothesis, as above, is the verdict of the design. What next? Does it mean that all is lost? Not at all. We can make another design and road test it again. Suppose we reflect that the quantum of increment in a pressure of nine units is quite adequate for the required purpose, then the new δ is nine instead of ten. The values of α and β can also be varied, meaning that the experimenter will accept the risk of making the α error and the β error at different probability levels. Let us say, for instance, that the new values of α and β chosen by the experimenter are both 0.05. Then, following the same steps of calculation, but with different numerical values, we have

$(5)_2.$ $$N = (1.645 + 1.645)^2 \times \frac{10^2}{9^2} = 13.35$$

or $N = 13$, and

$(6)_2.$ $$X = 250 + (10 \times 1.645) \div \sqrt{13} = 254.57$$

This means:

$(7)_2.$ After setting the temperature to 350°F, record the pressure readings over an extended period of time.

$(8)_2.$ Find $\bar{x}_1$, the mean of thirteen randomly selected pressure readings.

$(9)_2.$ Compare the new $\bar{x}_1$ with the new X, 254.57.

$(10)_2.$ Make the decision anew, with the same directions as before.

18.5.2 When σ Is Not Known

The process of design suffers a significant setback when σ, the standard deviation of the population, is not known. Having to deal with one more unknown—in fact a vital one, σ—the degree of reliability is considerably lowered. But when such situations are encountered, a method worked out by statisticians is available to bypass the setback. Broadly, this method consists of the following additional steps:

a. Finding N, the sample size, by two or more iterations.

b. Finding the degree of freedom.

c. Using t-distribution probability points in the second and later iterations.

d. Calculating a sample (substitute) standard deviation to be used in place of the unknown standard deviation.

Situation Set 2

All the other situations are exactly the same as in Example 1 with the only difference that σ is not known. The procedural steps are worked out, including the extra steps made necessary because of this difference.

Example 3:

1. σ is not known.

2. State the hypotheses:

   H_0: $\mu_1 = \mu_0$

   H_1: $\mu_1 = \mu_0 + 10$

This is one sided.

3. Choose the risk factors.

   $\alpha = 0.05$ and $\beta = 0.10$

4. Go to Table 18.2; look for U_α and U_β.

   For $\alpha = 0.05$, $U_\alpha = 1.645$

   For $\beta = 0.10$, $U_\beta = 1.282$

5. Compute N, the sample size.

   $N = (U_\alpha + U_\beta) \times \sigma^2 / \delta^2$

....

The only term now unknown in the RHS (right hand side) of the equation is σ, and that is what makes the difference in the steps and equations that follow.

....

Suppose σ is specified in terms of δ (it is frequently done so), for instance, $\sigma = (1.1) \times \delta$. If it is not so specified, the experimenter has the additional responsibility of deciding on a similar specification at his own discretion. Then,

$(5)_2$. $N = (1.645 + 1.282)^2 \times [(1.1\delta)^2 \div (1.0\delta)^2] = 10.37$

5b. Find the degree of freedom, given by

$\Phi = N - 1$

$= 10.37 - 1 = 9.37$

....

Note: The use of probability points for α and β, taken from Table 18.2, was good only for the first iteration, done above, for calculating N. For the second iteration, the probability points, in place of α and β, should be taken from Tables 18.4 and 18.5, which present *t-distribution points*, the values designated as t_α and t_β. Table 18.4 provides values of t_α and t_β when the alternate hyposthesis stated in step 2 is "one (single) sided." Table 18.5 provides those values when the alternate hypothesis is "two double) sided." Also, note that the t-distribution points are given in Tables 18.4 and 18.5, corresponding to Φ, the *degree of freedom*, a new parameter. These are the major differences.

Probability Points of t-Distribution; Two Sided; σ Unknown

$(4)_2$. Look up t-distribution points in Table 18.4, corresponding to $\Phi = 9.37$ (interpolating):

$t_\alpha = t_{0.05} = 1.82$; $t_\beta = t_{0.10} = 1.0$

$(5)_3$. Compute the sample size again, this time using t-distribution points in place of U_α and U_β; it is given by

$N_t = (t_\alpha + t_\beta)^2 \times \sigma^2/\delta^2$

$= (1.82 + 1.37)^2 \times (1.1\text{d})^2/(1.0\text{d})^2$

$= 12.31$

....

Table 18.4 *Probability Points of t-Distribution: One-Sided; σ Unknown*

φ	0.005	0.01	0.025	0.05	0.10	0.20	0.30
1	63.66	31.82	12.71	6.31	3.08	1.38	0.73
2	9.93	6.97	4.30	2.92	1.89	1.06	0.62
3	5.84	4.54	3.18	2.35	1.64	0.98	0.58
4	4.60	3.75	2.78	2.13	1.53	0.94	0.57
5	4.03	3.37	2.57	2.02	1.48	0.92	0.56
6	3.71	3.14	2.45	1.94	1.44	0.91	0.56
7	3.50	3.00	2.37	1.90	1.42	0.90	0.55
8	3.36	2.90	2.31	1.86	1.40	0.90	0.55
9	3.25	2.82	2.26	1.83	1.38	0.89	0.54
10	3.17	2. 76	2.23	1.81	1.37	0.89	0.54
15	2.95	2.60	2.13	1.75	1.34	0.87	0.54
20	2.85	2.53	2.09	1.73	1.33	0.86	0.53
25	2.79	2.49	2.06	1.71	1.32	0.86	0.53
30	2.75	2.46	2.04	1.70	1.31	0.85	0.53
60	2.66	2.39	2.00	1.67	1.30	0.85	0.53
120	2.62	2.36	1.98	1.66	1.29	0.85	0.53
μ	2.58	2.33	1.96	1.65	1.28	0.84	0.52

(Source: Wadsworth, Inc., 1981,William J.Diamond, *Practical Experiment Designs for Engineers and Scientists*)

Note: The above value, 12.31, for sample size is the result of the third iteration. A fourth iteration is known to be a closer approximation, which can be obtained by repeating steps 5 to 4 to 5 again, using the corresponding *t*-points.

For instance,

$(5b)_2$. Find the degree of freedom, given by

$\Phi_3 = N_t - 1$

$= 12.31 - 1 = 11.31$

$(4)_3$. Look for *t*-points in Table 18.4, corresponding to Φ = 11.31:

$(t_\alpha)_3$ 1.80; $(t_\beta)_3 = 1.36$

Table 18.5 *Probability Points of t-Distribution; Two-Sided; σ Unknown*

φ	0.005	0.01	0.025	0.05	0.10	0.20	0.30
1	127.00	63.70	31.82	12.71	6.31	3.08	1.96
2	14.10	9.93	6.97	4.30	2.92	1.89	1.39
3	7.45	5.84	4.54	3.18	2.35	1.64	1.25
4	5.60	4.60	3.75	2.78	2.13	1.53	1.19
5	4.77	4.03	3.37	2.57	2.02	1.48	1.16
10	3.58	3.17	2.76	2.23	1.81	1.37	1.09
15	3.29	2.95	2.60	2.13	1.75	1.34	1.07
20	3.15	2.85	2.53	2.09	1.73	1.33	1.06
25	3.08	2.79	2.49	2.06	1.71	1.32	1.06
30	3.03	2.75	2.46	2.04	1.70	1.31	1.05
60	2.91	2.66	2.39	2.00	1.67	1.30	1.05
120	2.86	2.62	2.36	1.98	1.66	1.29	1.05
μ	2.81	2.58	2.33	1.96	1.65	1.28	1.04

(Source: Wadsworth, Inc., 1981,William J.Diamond, *Practical Experiment Designs for Engineers and Scientists*)

$(5)_4$. Compute sample size again:

$$(N_t)_3 = (1.80 + 1.36)^2 \times [(1.1\ \delta)^2 \div (1.0\delta)^2] = 12.08$$

The difference in sample size as computed by the fourth iteration is not significantly different from that by the third iteration; thus, usually the third iteration is considered quite adequate in most cases.

Now that the preceding procedural steps have been carried out with $N = 12$, the experiment is ready to test drive.

....

7. Set the temperature at 350°F; after a reasonable length of time, record several pressure readings, x_1, over an extended period.

8. Find the mean of twelve randomly selected pressure readings, $\overline{x}_1$.

8a. Find the standard deviation of these readings, given by

$$S = \sqrt{\Sigma \frac{(x_1 - \overline{x}_1)^2}{N_t - 1}}$$

Note: S, calculated as above, will now be substituted in place of the standard deviation, S, which is not known. Now, we are back on track, as if σ were known.

---

6. Using S, the estimated standard deviation, compute the criterion value, given by

$$X = \mu_0 + t_\alpha \times \frac{S}{\sqrt{N_t}}$$

Note: Compare this equation with the parallel equation for case A (σ known). Notice t_a in place of U_a, and N_t in place of N.

---

9. Compare $\overline{X}_1$ with X.

10. Make the decision:

    If $\overline{X}_1 > X$, accept the alternate hypothesis,

    H_1: $\mu_1 = \mu_0 + 10$, meaning there is a 95 percent possibility that the pressure will increase to 260 psi.

If, instead, $\overline{X}_1 < X$, accept the null hypothesis, H_0: $\mu_1 = \mu_0$, meaning there is a 90 percent possibility that the pressure will remain unchanged.

18.6 Other Situation Sets

Thus far we have dealt in detail with only, perhaps, the two most used situation sets. We have seen the difference of requiring the additional steps in the procedure for situations in which σ is not known. This difference applies to all other situation sets as well. We will now visit a few more situation sets, some significantly different from the above two; for each we will mention the difference in the formulas used for computing N, the sample size X, the criterion value, and, further, the rules for accepting or rejecting the hypotheses.

Situation Set 3

All the situations in this set are exactly the same as in Situation Set 1, with the only difference that the alternate hypothesis here is

$$H_\alpha: \mu_1 < \mu_0$$

Procedure:

The steps are like those for Situation Set 1; the difference is only in step (9), the criteria for accepting one of the hypotheses.

Calculate the criterion number:

6. $X = \mu_0 + \sigma \times \frac{U_\alpha}{\sqrt{N}}$

Then, after getting $\overline{x}_1$, the mean of the improved dependent variable,

10. Make the decision:

If $\overline{x}_1 \leq X$, accept the alternate hypothesis, H_α: $\mu_1 < \mu_0$ with confidence of $(1 - \alpha)$ 100 percent or higher.

If $\overline{x}_1 > X$, accept the null hypothesis, H_0: $\mu_1 = \mu_0$.

Situation Set 4

Example 4:

This is a test for the simple comparison of two populations, A and B: no experiment is involved.

This, unlike Situation Sets 1 and 2, is two sided, meaning it can be either $\mu_1 > \mu_0$ or $\mu_1 < \mu_0$.

μ_0 and μ_1 are both known.

1. σ, the standard deviations of the measurements used for both μ_0 and μ_1, is known.

5. N_A and N_B, the number of samples from the two populations, both are known.

Procedure:

2. State the hypotheses:

   Alternate hypothesis: H_α: $\mu_1 \neq \mu_0$

3. Choose ρ (similar to α and β), the risk factor. This, like in Situation Set 1, is a subjective judgment made by the experimenter.

4. Go to Table 18.3, which gives the probability points U for the above set of situations.

   Corresponding to ρ chosen, find $U\rho$.

6. Calculate the criterion number.

$$[X_A - X_B] = U_\rho \times \sigma \times \sqrt{[\frac{1}{N_A} + \frac{1}{N_B}]}$$

10. Follow the rules:

    a. If $[X_A - X_B]$ is a positive number, accept the alternate hypothesis, H_α^1: $\mu_A > \mu_B$, with a confidence of $(1 - \alpha/2) \times 100$ percent or more.

    b. If $[X_A - X_B]$ is a negative number, accept the alternate hypothesis, H^2_a: $\mu_A < \mu_B$, with a confidence of $(1 - \alpha/2) \times 100$ percent or more.

    c. If $[X_A - X_B]$ is zero, accept the null hypothesis, H_0: $\mu_A = \mu_B$, with a confidence of $(1 - \beta) \times 100$ percent or more.

Situation Set 5

All conditions in this set are exactly like those in Situation Set 4, with the only exception that $\sigma_1 \neq \sigma_2$, though both σ_1 and σ_2 are known.

6. Then, the criterion number is given by

$$[XA - XB] = U_{\alpha}\sqrt{[\frac{\sigma_A^2}{N_A} + \frac{\sigma_B^2}{N_B}]}$$

The decisions on accepting and rejecting the hypothesis, based on the criterion number, are the same as in Situation Set 4.

Situation Set 6

The same specimen (or a part or a person) is measured for a given property (or capacity or performance) (1) before and (2) after improvement. Several such specimens, together, constitute the subjects for the experiment. Examples of such situations are (1) the effect of hardening heat treatment on a given alloy, (2) a new drug being tested for lowering the blood pressure of hypertension patients, and (3) the kick length of a soccer ball tested for a group of players before and after they take a miracle drink. The difference in property (or capacity or performance), symbolized by μ_0, is the population under consideration.

Example 5:

Each of the twenty-seven players on a high school soccer team are made to kick the soccer ball three times; the longest kick of each player is recorded as an element in the first sample. Each player is then given a measured quantity of a miracle drink. Ten minutes after taking the drink, the same twenty-seven players are made to repeat the performance of kicking the ball; the second sample, with a similar procedure as for the first, is collected from this repetition.

Procedure:

1. It is considered reasonable in such situations to assume the numerical value of σ to be the same as that of δ; hence, the question as to whether σ is known or unknown disappears.
2. State the hypotheses:

   H_0: $\mu_1 = \mu_0$

   H_α: $\mu_1 > \mu_0$
3. Choose α and β.
4. Look up for U_α and U_β (one sided, Table 18.2).

5. Compute N_1, the sample size:

$N_1 = (U_\alpha + U_\beta)^2\ \sigma^2/\delta^2$

$= (U_\alpha + U_\beta)^2 \times (1\delta)^2/\delta^2 = (U_\alpha + U_\beta)^2$

5a. Find the degree of freedom, given by F = $(N_1 - 1)$.

$(3)_2$. Look for t-values for Φ (one sided, Table 18.4).

Second iteration:

$(5)_2$ $N_2 = (t_\alpha + t_\beta)^2$

7. Experiment with N_2 sample items.

8. Find $(X_\beta - X_\alpha)$, the difference between X_β, the lengths of the kick after taking the miracle drink, and X_α,the corresponding length before taking the drink, for each one of the randomly selected N_2 players.

9. Find $\overline{x}_1$, the mean of the N_2 of such $(X_\beta - X_\alpha)$ values:

$\overline{x}_1 = \sum(X_\beta - X_\alpha) \div N_2$

$(1)_2$. Find s, the estimated standard deviation for N_2 measures of $(X_b - X_a)$:

$$s = \sqrt{\frac{\Sigma[(x_b - x_a) - \overline{x}]^2}{N_2 - 1}}$$

$(5a)_2$. Find the degree of freedom, second iteration: $\Phi_2 = N_2 - 1$.

3. Find t_α for Φ_2 (one sided, Table 18.4).

6. Compute the criterion value, $X = \mu_0 \div \dfrac{t_2 \times s}{\sqrt{N}}$

9. Compare $\overline{x}$, and X, and

10. Make the decsion:

9 and 10. Compare $\overline{x}_1$ and X, and make the decision:

If $\overline{x}_1 > X$, accept the alternate hypothesis with $(1 - \alpha) \times 100$ percent confidence.

If $\overline{x}_1 < X$, accept the null hypothesis.

18.7 Operating Characteristic Curve

A given set of experimental situations can be represented in a graphical form, referred to as the *operating characteristic curve* for that set. To show how it is derived, we will revisit the design for experimental Situation Set 1.

H_0: $\mu_1 = \mu_0$ (average pressure, 250 psi)

(δ = 10; enhanced pressure, 260 psi)

In road testing the design, suppose δ falls short of ten. It is quite possible that the experimenter will be satisfied with an increment of 8 psi instead, but he likes to know what the probability of getting 258 psi is in the road test. And, what are the probabilities of getting 257, 256, 255, and so forth, just in case he can settle down to one of those? A graphical relation, known as the *operating characteristic curve*, which can help the experimenter forecast answers to such questions, even before the road test, is our aim.

Now, we will look at the two formulas central in the design:

$$N = (U_\alpha + U_\beta) \times \frac{\sigma^2}{\delta^2} \tag{18.1}$$

$$X = \mu_o + \frac{\sigma \times U_\alpha}{\sqrt{N}} \tag{18.2}$$

Between these two relations, the factors involved are μ_0, σ, δ, U_α, U_β,

$\overline{X}_1$, the mean of the enhanced pressure values, yet to be obtained from road testing the design, symbolized as μ_1, and,

α and β, the risk factors, which, in turn, lead us respectively to U_α and U_β (Table 18.2).

With these factors, the experimental Situation Set 1, in terms of numbers, reduces to

μ_0= 250

μ_1 = 260

$\sigma = 10$

$\delta = 10$

$\alpha = 0.05$

$\beta = 0.10$

The enhanced pressures, 257, 256, and so forth, are simply the values of μ_1, its value becoming 260 with $\delta = 10$.

The operating characteristic curve, as mentioned earlier, is a graphical relation of probabilities for μ_1 varying between 250 psi at one end and 260 psi at the other. The pressure 250 psi is given as an established fact for the design. It means the possibility of getting 250 psi is 100 percent. But because we have to contend with probabilities, not certainties, the possibility is less than 100 percent. What is the probability? The meaning of having chosen $\alpha = 0.05$ is as follows: Although the pressure of 250 psi is an established fact, rejecting the null hypothesis amounts to asserting that there is bound to be an improvement in the pressure. This will be a statement in certainty, not probability. Conforming to probability, all that we can say is that there is a very high probability that this could happen. And that probability is $(1 - \alpha) = (1 - 0.05) = 0.95$. If we plot pressure readings on the *X*-axis and probability on the *Y*-axis, there is a point with coordinates $X = 250$, $Y = 0.95$. And, though we are not sure that there is an enhancement of 10 points in pressure ($\delta = 10$, $\mu_1 = 260$), there is some possibility that this may happen; the probability of that is 0.10. Now we have the second point on the graph with coordinates $X = 260$ and $Y = 0.10$. We have probability numbers corresponding to the μ_1 (enhanced pressures) = 250 and 260 psi. What is the probability number for an intermediate pressure, for example 256 ($\delta = 6$) psi? To answer this, we will go to (18.1) above:

$$N = (U_\alpha + U_\beta) \times \frac{\sigma^2}{\delta^2}$$

, which can be simplified to

$$U_\beta = \sqrt{N}\frac{\delta}{\sigma} - U_\alpha \tag{18.3}$$

In this formula, all the factors except δ remaining unchanged, we can calculate the value of U_β for $\delta = 6$.

$$U_\beta = \left(\sqrt{9}\times\frac{6}{9}\right) - 1.645 \text{ (for } \alpha = 0.05,\ U_\alpha = 1.645)$$

= 0.355 (assumption: all the road tests will be done with nine samples)

Now, going to Table 18.2, we can find (extrapolating), corresponding to $U_\beta = 0.355$, the value of β to be 0.362.

This is the third point with coordinates X = 256 and Y = 0.362 on the probability curve. Calculations along similar lines for $\delta = 8$ give us another point: $X = 258$, $Y = 0.154$. A bell-shaped (because it is normal distribution), smooth curve, drawn through these four points, gives the operating characteristic for this set of experimental conditions. For any given value of enhanced pressure in the range 250 to 260 (δ 0 – 10), for example, 254, the experimenter is able to find from this curve the probability of attaining it.

We note that for drawing this curve, we used (18.1) and (18.3), but made the assumption that N, the sample size, remained constant for all the road tests yet to be done. Instead, (18.1) can be used to find the sample size with varying increments in the dependent variable. For instance, with other data unchanged, the sample size, N, for $\delta = 4$ is given as

$$N = (1.645 + 1.282)^2 \times 10^2/4^2 = 54$$

In the extreme case of $\delta = 1$, a similar calculation gives us N = 857, an ungainly number for the sample size, particularly if each numerical value in the sample requires a replication. We notice that the difference we are considering in the response—the independent variable in this case, the pressure—is only 1 in 251. The implication is that situations in which minute changes in the response need to be detected require very large samples. The smaller the difference to be detected, the larger the sample required.

Suppose we take the sample size to be fifty-four, but keep the other variable factors the same in (18.3). Then, corresponding to $\delta = 4$ and 2 (pressures 254 and 252), the values of β work out to be 0.0512 and 0.217, respectively. We can plot the points ($x = 4$, $y = 0.0512$) and ($x = 2$, $y = 0.217$). A bell-shaped, smooth curve passing through these points is good for sample size fifty-four only. Both these curves, one for $N = 9$ and another for $N = 54$, were obtained for conditions with $\alpha = 0.05$. Now, if δ is changed, the location and paths of the curves change too.

The curves, which show the probabilities of attaining, in steps, a given target for improvement, expressed in concrete terms of a problem, are often referred to as *power curves.* The same data may be expressed in general terms, namely: "Probability of accepting alternate hypothesis" on the Y-axis and "Increments of δ between μ_0 and μ_1" on the X-axis. Then, the curves are properly called the *operating characteristic curves.* Data from such curves can be applied to other problems with a comparable range of variables in experimental situations. Figure 18.1 shows these curves, alternatively to be called by either of the titles, depending on the way the X and the Y coordinates are expressed.

Finally, we take a closer look at (18.3), which is the basis for plotting the operating characteristics curve. It is a relation of probability U_β to four variables:

a. The sample size, N

b. The intended increment

c. U_α, hence α

d. The standard deviation, α, which is likely to be known

The first three of these variables are interdependent. For example, by increasing N and decreasing δ, the value of U_β can be kept the same. Conversely, the value of U_β can be changed by altering any one or more of N, δ, and U_α (hence α). Though these variables are situation specific, they are open to the discretion of the experimenter. A large number of graphical relations, connecting these variables to the probability of attaining the targeted improvements by experiments, are can be found in *Biometrika Tables for Statisticians* Vol 2, by E. S. Pearson and H. O. Hartley, Cambridge University Press, Cambridge 1972. The reader is advised to refer to a standard text on "Design of Experiments,"

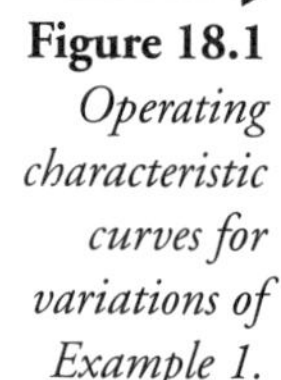

Figure 18.1 *Operating characteristic curves for variations of Example 1.*

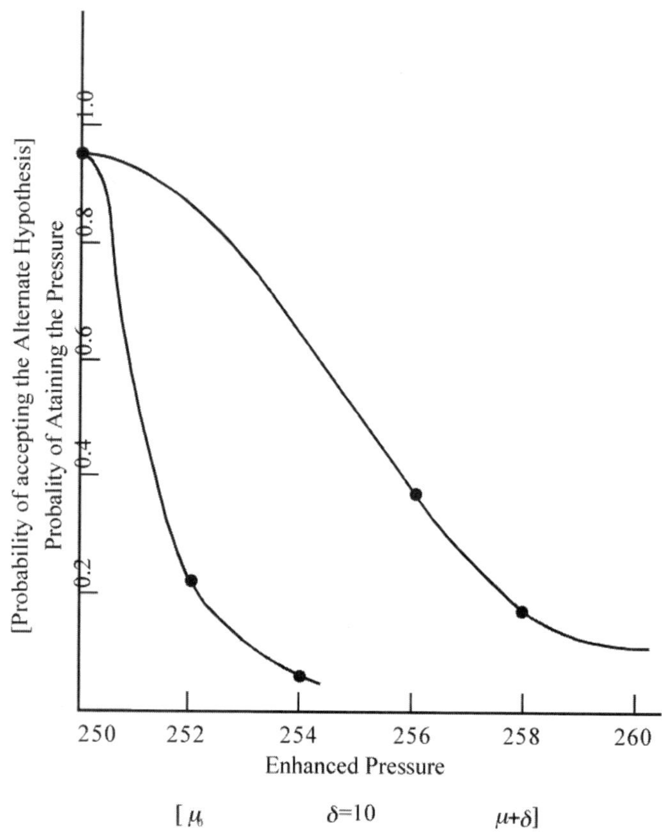

many of which (e.g., D. C. Montgomery, *Design and Analysis of Experiments*, 5th ed. New York: John Wiley and Sons, 2001) contain selected adaptations of such data; these can be applied, with some interpolation, to specific experimental situations.

18.8 Sequential Experimenting

As the reader should have noticed, the two key steps in designing the experiment in all the variations so far presented are (1) finding the number of items in the sample *N*, and (2) computing the criterion value, *X*, to compare it with the mean of the response output, $\overline{X}_1$. If the sample items are such things as specimens for hardness measurement or a group of players kicking a soccer ball, having all the *N* sample items together, in terms of either money or time, is not a problem. Instead, suppose the sample items to be tested are rockets of a novel design. It is prohibitively expensive both in terms of money and time to collect all *N* items, each with different levels of variable factors, and to fire them simultaneously, possibly at the end of a long waiting period. Assembling each item may take a few weeks, if not longer. In such cases, experimenting with one item at a time, followed by a long preparation time before the second experiment is to be done, is reasonable. Also, at the end of each test, the data is subjected to the test of significance somewhat on the same basis as, but with different details from, that of the experiment. As soon as the data passes the test of significance, the project of experimentation is terminated. Most often, it happens that the number of sequential tests necessary is much smaller than the number *N*, evaluated by computation. The first three steps of the procedure shown below are the same as in the previous designs, but the subsequent steps are specially formulated (by statisticians) for the situation. An example is illustrated using typical hypothetical numbers.

Example 6:

This experiment was for a metallurgical enterprise. A steel part was shipped out to a possible customer, who proposed to buy this component if it was approved as a vendor item. The parts were hardened by carbon diffusion. This process, with the present setup, required about fifteen hours of soaking in the furnace; 26 percent of the parts were rejected by the possible customer because they failed to show the required hardness of 88 HRC (Rockwell Hardness on C Scale). The measurement of hardness by the enterprise before shipping was recorded as 90 HRC with a standard deviation of four. The in-house metallurgist proposed to make certain changes in the process so that the parts could meet the customer's requirement. The effect of one set of changes in the process could be tested for on consecutive days.

The sequential experiment begins at this point. As was shown earlier in this chapter, the procedural steps differ for the situation sets:

a. σ is known; $\mu_0 > \mu_1$.

b. σ is known; $\mu_0 \neq \mu_1$.

c. σ is not known; $\mu_0 > \mu_1$.

Procedure:

The present experiment conforms to set α.

1. σ is known to be four.
2. State the hypotheses:

   H_0: $\mu_1 = \mu_0 = 88$

   H_α: $\mu_1 > \mu_0$; this is one sided.
3. Choose $\alpha = 0.05$, $\beta = 0.025$, $\delta = 4$.
4. Find in Table 18.2, $U_\alpha = 1.645$, $U_b = 1.960$.
5. Calculate $N = (1.645 + 1.960)^2 \times \dfrac{4^2}{4^2} = 13.$

....

Up to this point, the procedural steps are the same as for Situation Set 1. Steps from now on are different for sequential experiments. The numbers here onwards assigned to steps do not correspond to the same functions as in Examples 1-5.

....

Compute

$$U = \frac{\delta}{2} + \frac{a \times \sigma^2}{N \times \delta}$$

and

$$L = \frac{\delta}{2} - \frac{b \times \sigma^2}{N \times \delta}$$

in which

$$a = \ln\,[(1 - \beta) \div \alpha] = \ln\,[(1 - 0.025) \div 0.05] = 2.97$$

and

$$b = \ln\,[(1 - \alpha) \div \beta] = \ln\,[(1 - 0.05) \div 0.025] = 3.64$$

U and L for the present situation set simplify to

$$U = \frac{4}{2} + \frac{2.97 \times 4^2}{N \times 4} = 2 + \frac{11.88}{N}$$

and

$$L = \frac{4}{2} - \frac{3.64 \times 4^2}{N \times 4} = 2 - \frac{14.56}{N}$$

6. With formulas for U (standing for the upper) and L (standing for the lower) known, plot two lines on the same graph, corresponding to $N = 1, 2, 3, 4$ . . . as shown in Figure 18.2.

Figure 18.2 *Sequential experimenting.*

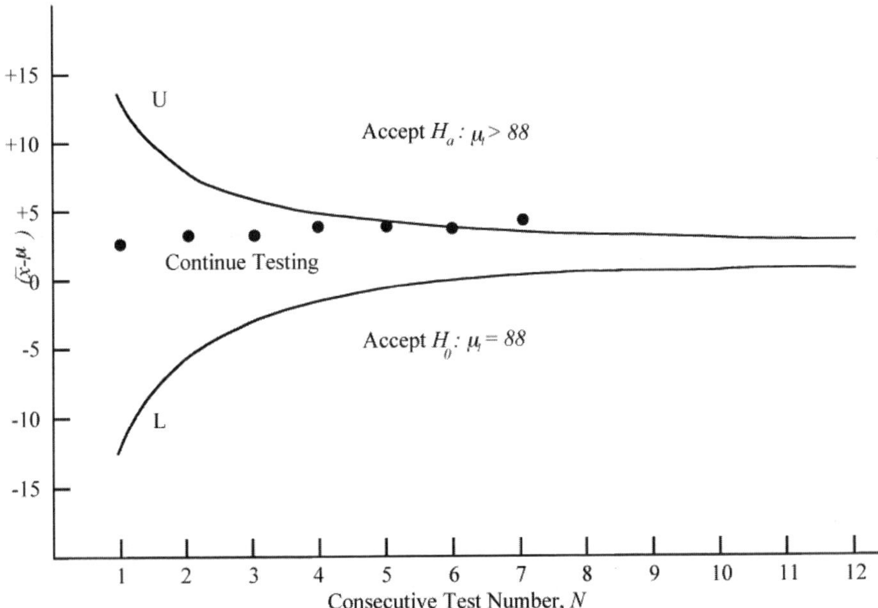

Note: These lines serve the same purpose as the criterion number, X, in Situation Set 1. After plotting these lines, one can notice that the graph area is divided into three fields. Mark these areas as shown in Figure 18.2: (1) accept H_α, (2) continue testing, and (3) Accept H_0. The rest of the procedure is nearly the same as in Situation Set 1, except that in place of *criterion number*, now we have *criterion lines*.

7. Conduct the first experiment ($N = 1$), and get the output value, χ_1. This is the first experiment; now $X = \chi_1$.
8. Find $(X - \mu_0)$.
9. Plot this value as ordinate in the graph corresponding to abscissa $N = 1$.

Note: The location of this point, meaning the particular field of the graph in which it falls, directs the decision. If it falls in either area "accept H_α" or "accept H_0," that is the end of the experimentation. If it falls in neither, it will be in the area "continue testing."

Then,

10. Prepare the second specimen and perform the second experiment. Let us say the output of this experiment is χ_2. Then,

$X = (\chi_1 + \chi_2) \div 2$

Find $(X - \mu_0)$.

Plot this value in the graph corresponding to $N = 2$.

Again, depending on the location of this point, make a decision.

Continue, if need be, with $N = 3$, $N = 4$, and so on.

The very first time a plotted point derived as above falls outside the area "continue experiments," the decision is accepted as final, and the project of experimentation is terminated. If the point is located in "accept H_α," that is what the decision is; if it is "accept H_0," likewise, that is the decision.

The results of these steps, one for each $N = 1, 2, 3 \ldots$ are shown in Figure 18.2. We find that the point corresponding to the seventh test falls for the first time in the field "accept H_α." That is the end of the experiment. Whatever changes were made in the parameters at that step will then be followed as routine in heat-treating the parts. The required number of parts processed according to the new parameters will be sent to the customer, now with new hope for approval.

18.9 Concluding Remarks on the Procedures

Thus far in the procedure for designing the experiments, we have encountered the following major steps:

a. Expressing the specific purpose of the experiment in the form of two statements: (a) the null hypothesis, and (b) the alternate hypothesis, often, but not necessarily, in terms of numbers

b. Expressing the risks that are necessarily involved, in terms of probability numbers, in making the decision of accepting or rejecting either of the two hypotheses

c. Calculating the sample size, according to the given formulas, involving the risk factors and the situations specific to the experiment

d. Expressing, in the form of equations, the objective criteria, which serve as the reference for deciding to accept or reject either of the two hypotheses

These four steps of the design can be carried out when the standard deviation of the population is known. But, the logic in following steps 3 and 4 is not quite transparent to the reader. When σ is not known, the degree of uncertainty in the experiment increases. The decision-making process relative to accepting or rejecting the hypotheses, as well as the determination of the optimum sample size, is affected. Because of this additional uncertainty, the risk factors α and β cannot be taken from a normal distribution curve, but instead should be taken from the curves known as *t*-distribution curves. This, in turn, involves another factor, Φ, known as the degree of freedom, usually expressed by $\Phi = N - 1$. The "estimated value of σ" takes the place of σ (if it was known), and then follows another round of calculations for the sample size, the criterion value, and so on. The logic involved in all such procedures is practically opaque.

The reader is led through these procedures somewhat blindfolded. We will close this chapter with a note of reassurance to the reader. You are asked here, more than in any other chapter, to accept numbers, use equations, and follow procedures “blindly” without questioning why. The answer is, because of the nature of the statistical methods. The abundant tables and graphical data attached as appendixes to almost all the texts on statistics testify to this situation. The rationalizing—the logical—part is carried out by statisticians; others use statistics more or less blindly. The uneasy reader may reflect on the use of computers. How many of the users wait to know the logic underlying the many feats that they can perform with computers?

18.10 Bibliography

Anderson, David R., Dennis J. Sweeny, and Thomas A. Williams. *Introduction to Statistics: An Applications Approach*. St. Paul, MN: West Publishing Co., 1981.

Diamond, William J. *Practical Experimental Designs for Engineers and Scientist*. New York: John Wiley and Sons, 2001.

Winer, B. J. *Statistical Principles in Experimental Design*. New York: McGraw-Hill, 1971.

19

Statistical Inference from Experimental Data

Scientific knowledge . . . is not arrived at by applying some inductive inference procedure to antecedently collected data, but rather by what is often called "the method of hypothesis," i.e., by inventing hypotheses as tentative answers to a problem under study, and then subjecting these to empirical tests.

—*Carl G. Hempel*

In this last chapter of the book, interpretation of the raw data, the harvest of the experimentation, is described. The experimenter, in the course of his work, will have collected a large amount of data; these will be mostly in the form of numbers in quantitative research. The data will be sorted into many bunches of numbers. Based on these numbers, the experimenter will be expected to make generalizations. Such generalizations can be only reasoned guesses because, however large the numbers in each bunch may be, in the context of experimental research, each bunch is only a sample, not a population. Available statistical methods to take the guesswork out and, instead, assign a reasonable *degree of confidence*, confidence expressed in terms of probability, are described. Some relations between variables—independent and dependent—may be expressed as algebraic equations. The procedures required to formulate the equations, referred to as *regression analysis*, are outlined.

19.1 The Way to Inference

During the life span of an experiment, following the stages of conducting the experiment and collecting data, there comes a time when the experimenter, with his logbooks loaded with measurements, meter readings, and various counts of other kinds, will have to sit and ponder how reliable each bunch of observations (measurements, readings, or counts) is for answering his questions, and how such answers fit together in the purpose for his experiment. At this stage, he will be *analyzing the data*, the purpose of which is to make decisions, predictions, or both about unobserved events. We may ask, What are such unobserved events? The answer is, They include all things that are yet to happen in situations similar to the ones that are embodied in the experiment. The relevance of this answer becomes obvious if we recall that making generalizations is the major purpose of research. Based on what we observe happening, we feel emboldened to predict what may happen in future and elsewhere under similar circumstances.

The experimenter, while analyzing the data, finds himself in a situation wherein such concerns as the following are all behind him in time:

- Standardizing the material
- Randomizing relative to the material, the subject, the sequence of experiments, and other relevant conditions
- Calibrating all measuring devices involved
- Designing the experiment (1) to avoid nuisance factors and (2) to account for factors that have combined effects

All that is now on hand are several bunches of observations, each observation being either a yes-no response or, more often, a number. We further imagine that each number is the output of an experimental inquiry, in the form of a dependent variable, the system responding to an input, in the form of one or more independent variables. The bunch of observations we referred to consists of several such outputs recorded while replicating the experiment.

Our discussion here is restricted to statistical inference, which means the process of making generalization about populations, having at our disposal only samples. In the context of experimental research, each of what we called a "bunch of numbers" is a sample. If that is so, where is the population? It exists only as a possibility; the possibility of any person at any time in future, conducting a similar experiment and getting another bunch of numbers. Suppose the bunch of numbers we have on hand is 56.4, 56.4, 56.4, . . ., all the *n* elements of the set being 56.4. The *mean* of this set is 56.4 and the *standard deviation* is 0. Then, we project that the population, however big, will have for its mean 56.4, and for its standard deviation, 0. But this situation practically never occurs. Even if all possible controls are exercised in experimentation, individual numbers in the bunch (elements in the set) will have variation. And that is the reason why, and the occasion when, *statistical inference* is required. It consists of two parts: (1) trying to find statistical parameters of the population such as mean and standard deviation, and (2) formulating specific assertions, known in statistics as *hypotheses*, to test in statistical terms whether such assertions can be accepted or need to be rejected. The first part mentioned above is known as *estimation*, and the second as *testing of hypothesis*. It is necessary for the reader to distinguish further here between "hypothesis" as used in the logic of planning experiments, as was done in Chapter 18, and the meaning of the same word as used here in statistics and referring to the data collected from experiments. The scope of the latter, *testing of (statistical) hypothesis*, is rather limited. It is confined each time to deriving inferences from one set of numbers on hand, which is done in the following pages.

19.2 Estimation (From Sample Mean to Population Mean)

As pointed out in Chapter 18, the mean is the most significant statistic of a given set. The mean of the sample set *A*—56.4, 57.2, 55.9, 56.2, 55.6, 57.1, and 56.6—is 56.43. This is called the *sample mean*, $\bar{x}$, and when it is used as an estimator of the *population mean*, μ, it is referred to as the *point estimate*, meaning that it is not a range of μ values. The drawback to point estimation is that it does not, by itself, indicate how much confidence we can place in this number. Suppose 56.43 came out to be the mean of a set of fifty elements, instead of a set of only seven elements as above; intuitively, we would accept it with more confidence. The higher the number of elements in the sample, the higher the confidence, meaning the better the estimate.

We also want to look at the part played by another statistic, namely, the standard deviation, in deciding the confidence level that a sample mean commands in the analysis of results. Instead of the sample set *A* seen above, say we had a sample set *B* consisting of (as before) seven elements: 61.4, 60.2, 50.9, 61.3, 51.0, 57.6, and 56.2. The mean of this set is also 56.43, the same as before, but we can readily notice the difference: whereas in set *A*, the individual values were close to each other, in set *B* there is a wide variation. Between the two samples, intuitively, we place more confidence in *A* than in *B*. As is now familiar, the statistic that measures this variation is the *variance*, or standard deviation, which is simply the square root of the variance. In summary, we may say that, in terms of confidence, the lower the standard deviation of the sample, the better the estimation.

Thus, any sample, however large, cannot provide a sample mean that can be accepted with absolute confidence as equivalent to the population mean. Some error, known as the *sampling error*, and given by

$$\text{Sampling error} = (\overline{X} - \mu)$$

is always likely to occur. To avoid this drawback, another method of estimating the mean, μ, is often used; it is known as interval estimation.

19.2.1 Interval Estimation

In the ideal situation, where the sample statistic is equal to the population parameter (as in the particular case we discussed for the two means), the sample statistic is said to be an *unbiased estimator* of the corresponding population parameter. But in reality, this is never likely to happen. A sample statistic, for instance the mean, is something that we can access and evaluate. By making use of it, suppose we establish an interval within which we can be confident to a specified degree that the population mean will be located; this is the process of *interval estimation*. The degree of confidence is expressed in terms of probability. The interval is known as the *confidence interval*; the upper and lower limits are known as *confidence limits*.

In demonstrating the methods of estimating the confidence interval for the population mean, μ, we want to distinguish three cases:

1. A large random sample (more than thirty elements) is possible, and the standard deviation, σ, of the population is known.
2. The sample size is more than thirty, and σ is unknown.
3. The sample size is small (less than thirty elements), and σ is unknown.

In all three cases, we assume arbitrarily for the present demonstration that the degree of confidence required for m is 95 percent (or 0.95).

Case 1: Sample Size Is More Than Thirty; σ Is Known

From the sample, the sample mean, $\overline{\text{X}}$, can be readily calculated. By the definition of interval estimation, $\overline{\text{X}}$ is expected to be located between the lower and the upper limits of the population mean, μ. The statistical instruments we invoke to develop the relations are (1) the Central Limit Theorem, and (2) the standard normal distribution (see Chapter 17).

According to the Central Limit Theorem, if the sample size is reasonably large ($N \geq 30$), the distribution (of many values) of μ approximates to normal distribution with mean μ and standard deviation

$$\sigma_{\overline{\text{X}}} = \frac{\sigma}{\sqrt{N}} \; .$$

In the graphical relation of this distribution, the distance on the abscissa between μ and a particular value of $\overline{\text{X}}$ gives, in effect, the difference between the population mean, μ, and the sample mean, $\overline{\text{X}}$. Now, adapting the notation of the standard normal distribution, this difference can be represented as

$$z = \frac{\bar{x} - \mu}{\frac{\sigma}{\sqrt{N}}} \quad (19.1)$$

in which N is the number of elements in the sample (or simply the *sample size*).

It is known that the area under the curve of standard normal distribution is the numerical measure of probability, P, whose maximum value is 1.0, and further that 95 percent of the area under the curve lies between the z values of –1.96 and +1.96. These facts can be formulated as

$$P(-1.96 < z < +1.96) = 0.95 \quad (19.2)$$

And, the verbal interpretation of this is that there is a 0.95 probability that the random variable, z, lies between –1.96 and +1.96, or (in other words) that the confidence interval for the parameter μ is 95 percent.

Combining (19.1) and (19.2), we get

$$[(\bar{x} - 1.96 \times \frac{\sigma}{\sqrt{N}}) < \mu < (\bar{x} + 1.96 \times \frac{\sigma}{\sqrt{N}})] = 0.95 \quad (19.3)$$

The verbal interpretation of this is that there is a 0.95 probability that μ lies between

$$(\bar{x} - 1.96 \times \frac{\sigma}{\sqrt{N}}) \text{ and } (\bar{x} + 1.96 \times \frac{\sigma}{\sqrt{N}})$$

In other words, the confidence interval for the parameter μ, between the above two limits, is 95 percent.

Example with hypothetical numbers:

Sample size, $N = 64$

Sample mean, $\bar{x}$ (calculated) = 26.4

Standard deviation, σ (known) = 4

Degree of confidence, selected = 0.95 (95 percent)

Using (19.3)

$$26.4 - 1.96 \times \frac{4}{\sqrt{64}} < \mu < 26.4 + 1.96 \times \frac{4}{\sqrt{64}}$$

$$= 26.4 - 0.98 < \mu < 26.4 + 0.98$$

$$= 25.42 < \mu < 27.38$$

By this, we understand that the true value of μ, the population mean, lies somewhere between 25.42 and 27.38 and that we can be 95 percent confident in accepting this.

Application:

The situation in which the standard deviation of the population is known but its mean is unknown, hence, needs to be estimated, is not very likely to be encountered with the data obtained from laboratory experiments in science or engineering. On the other hand, relative to data obtained from surveys in a national census, public statistics, psychology, medicine, education, and so forth, the situation as mentioned above is likely to occur. For instance, the standard deviation in the height of adult males in the United States established last year or the year before is likely to hold good this year as well. Also, in surveys of this kind, the sample size is usually large, and the assumption made to this effect is justified.

19.2.2 Variations in Confidence Interval

In the foregoing discussion and example, we arbitrarily chose the level of confidence to be 95 percent. It is possible, for instance, to choose it to be 99 percent (or any other number). Supposing it is 99 percent, the change required is as follows: Going to the table of standard normal distribution, we select the z value such that the area under one half of the curve (because the curve is symmetrical) is $0.99 \div 2 = 0.495$. That is found to be 2.575. Using the same data as in the above example, the new relation is

$$26.4 - 2.575 \ \frac{4}{\sqrt{64}} < \mu < 26.4 + 2.575 \ \frac{4}{\sqrt{64}}$$

$$= 26.4 - 1.287 < \mu < 26.4 + 0.287$$

$$= 25.11 < \mu < 27.69$$

We notice that if we require a higher degree of confidence (99 percent instead of 95 percent), we end up with a wider confidence interval. It is desirable to have a combination of a higher degree of confidence and a narrower confidence interval. To look for the variables that influence these two somewhat conflicting outcomes, we go to (19.3) and note the following: Retaining a higher degree of confidence, to narrow the confidence interval, we need to increase the factor to the left of μ and decrease the factor to the right of μ. $\bar{x}$ being common to both sides, and σ being known and fixed, the only changeable factor is $\sqrt{N}$, or N, the sample size. To the extent that N, the number of elements in the sample, is higher, the lower limit of the confidence range (on left) increases and the upper limit (on the right) decreases, both together rendering the confidence range narrower. This confirms what we mentioned earlier—that larger samples should give us more confidence in estimating the population parameters. Incidentally, also note that if σ, the standard deviation, were given to be lower (say two instead of four in the example we considered), the effect would have been similar to having higher N, namely, rendering the confidence interval narrower.

Case 2: Sample Size Is More Than Thirty; s_x Is Unknown

Standard deviation of the sample, s_x, is computed and substituted in place of σ in (19.3):

$$(\bar{x} - 1.96 \times \frac{s_x}{\sqrt{N}}) < \mu < (\bar{x} + 1.96 \times \frac{s_x}{\sqrt{N}}) \qquad (19.4)$$

The result of the procedure is the same as in Case 1.

Application:

This is more realistic than Case 1 and suitable to laboratory experiments.

Case 3: Sample Size Is Small (Less Than Thirty); S_x Is Unknown

Like in Case 2, S_x is computed and substituted in place of σ in (19.3). But that is not all; other changes follow. In place of the inequality part in (19.3) (contained within parenthesis), we need to use a modified format, namely,

$$\bar{x} - t_{0.025} \times \frac{s_x}{\sqrt{N}} < \mu < \bar{x} + t_{0.025} \times \frac{s_x}{\sqrt{N}} \tag{19.5}$$

wherein we find $t_{0.025}$ replacing 1.96, because instead of the standard normal distribution curve, we are required to use here what is known as the *t-distribution curve*. The area under this curve between the limits $-t_{0.025}$ and $+t_{0.025}$, like before, is 0.95, thus is the measure of confidence level required. The "0.025" in "$t_{0.025}$" is derived as (1 – 0.95) ÷ 2 = 0.025: this is because the *t*-distribution curve, like the standard normal distribution curve, is symmetrical and "two tailed." (If, instead, the confidence level required is 0.99, we need to use $t_{\frac{1-0.99}{2}} = t_{0.005}$). At this point, we go to the *t*-distribution table (Table 18.4). Before then, we need to make one more decision relative to Φ, the degree of freedom. Bypassing the logic for the statistics involved, we take the degree of freedom, as was done in Chapter 18, to be the number of elements in the sample, minus one. If the sample strength is known to be ten, for instance, the degree of freedom is taken to be (10 – 1) = 9. Now, corresponding to degree of freedom, nine (in the first column) and $t_{0.025}$ (in the third column), we have the *t* value 2.262.

Substituting in (19.5), 2.262 in place of $t_{0.025}$, $\bar{x}$ and s_x calculated for the sample, nine in place of *N*, we can compute the numerical values of the lower and the upper limit of the required parameter, namely, the population mean, μ. We can accept it with 95 percent confidence.

Application:

This method is suitable for most laboratory experiments in science and engineering, except, perhaps, the *control-subject experiments* in life science, agriculture, and medicine, wherein case studies in excess of thirty subjects are quite likely.

19.2.3 Interval Estimation of Other Parameters

Though mean is the population parameter most often sought, there are occasions when other parameters, particularly the standard deviation of a (real or hypothetical) population, are required. In manufacturing quality control, for example, where closer tolerances mean better quality, reducing standard deviation is always strived for. Following the same method of invoking the Central

Limit Theorem, the notation of standard normal distribution, we can work out the formula for the confidence interval for the standard deviation. As in the case of means, methods are different, as the case might be, for a large or small sample. The inequalities, one for the lower and another for the upper limit of confidence of a given probability value (that is, degree of confidence), have a similar format but with different N-related constants, where N is the sample size. For detailed procedures relative to these aspects of the estimation of any other parameter, the reader should refer to a standard text on statistics.

19.3 Testing of Hypothesis

Testing of statistical hypothesis has several applications, almost wherever statistics is applicable. In this book, however, we shall confine the discussion to experimental research, considering only one obvious and hypothetical case of application; this will show the way to other possible applications with minor modifications as required.

Let us imagine that an experiment has been conducted previously. If that experiment is replicated elsewhere at a later date, with no change in the independent variables, instrumentation, and material involved, we expect the value of the dependent variable to be the same as in the previous experiment. Suppose the value of the dependent variable was previously thirty. We further assume that thirty was an ideal minimum number to confirm a theory. If the corresponding value in the present experiment comes out to be twenty-nine, should the theory be rejected? In this situation, thirty, the number we made reference to, is in effect, the hypothesis. And, the statistical consideration and the procedures involved in rejecting or accepting the theory in question constitute the testing of the hypothesis.

Example:

This experiment consists of a student trying to confirm the results of a previous Ph.D. dissertation. The professor who has guided the Ph.D. work has recently come to doubt the result of that work; the reputation of the previous student (hence, his own) is being questioned. The Ph.D. work is supposed to have confirmed a theory developed by the professor. The confirmation depended on a certain minimum number, which was the outcome, as the dependent variable, of the experiment done with the specific independent variables in the experiment. There is a statement in the Ph.D. thesis to the effect that the mean of a large number of replications gave the value of the dependent variable as 53.2, with a standard deviation of 1.4. Thus, 53.2 is the number required to confirm the theory.

Another student of the same professor is the present experimenter. As far as this student can manage, he has reproduced the conditions of the Ph.D. experiments. His numbers are 51.5, 54.3, 53.1, 52.0, 53.6, 52.7, and 51.9 (we will suppress the units, millivolts, for convenience.) The mean of these readings, 52.7, is less than 53.2, the minimum expected. The present student is inclined to reject the previous confirmation of the theory, to that extent claiming credit for himself. The professor, having previously used the statistical methods, is not in such a hurry. Testing of the hypothesis starts here; the number 53.2 is at the core of the hypothesis.

In the following hypothesis, μ_0 represents 53.2, and μ_1 represents the mean of the values of the dependent variable from the current experiments. We state the hypothesis for this case somewhat differently from how we did before.

The null hypothesis, H_0: $\mu_1 \geq \mu_0$ (53.2), means, if μ_1 is equal to or greater than 53.2, there is no need to doubt the previous experiment; action is not necessary.

The alternate hypothesis, H_a: $\mu_1 < \mu_0$(52.3), means, if μ_1 is less than 53.2, doubt is valid, and action is necessary.

"Action" in both cases constitutes rejecting the theory based on the confirmation by previous experiment.

The fact that the mean of the present seven values is less than 53.2 is not statistically important because this is the sample mean, $\bar{x}$ (not the population mean), and that too calculated from a small sample. Further, dealing with probability, not certainty, we need to know how much lower than 53.2 the value of $\bar{x}$ can be to safely reject H_0. This situation calls for a criterion value to be compared with $\bar{x}$; we will designate such a value as *X*. Now the decision rules can be formulated as

- Accept H_0 if $\overline{x} \geq X$.
- Reject H_0 if $H_0 < X$.

The next steps are

1. To gct a morc dcpendable $\overline{x}$, meaning more replications, preferably more than thirty
2. To find by statistical analysis a "rcasonable" value of X

Assuming that the present experimenter is engaged in performing more replications, together amounting to 32(N), we will attend to the statistical part, step (2), below:

Applying the Central Limit Theorem (see Chapter 17), we know that the population mean, μ_0, is given by the mean of (many) $\overline{x}$ values and that the standard deviation of the $\overline{x}$'s, σ_x, is given by $\frac{\sigma}{\sqrt{N}}$, in which σ is the population mean.

In our case

$$\sigma_x = \frac{1.4}{\sqrt{32}} = 0.25$$

Now the probability of rejecting H_0 needs to be addressed. Being conscious that his reputation is at stake, the professor is likely to be very conservative. We assume that his decision is to allow only a 1 percent chance of ejecting H_0, meaning $\alpha = 0.01$. Using the notation of the standard normal distribution (see Chapter 17), we can now write the value of X:

$$X = 53.2 - Z_{0.01} \times \sigma_{\overline{x}}$$

Going to the table of standard normal distribution, we find the z value for the probability of $0.5 - 0.01 = 0.49$ to be 2.33. Substituting 2.33 for $Z_{0.01}$ and 0.25 for $\sigma_{\overline{x}}$, we can find X.

$$X = 53.2 - 2.33 \times 0.25 = 52.6.$$

Now we can state the decision in terms of concrete numbers: Accept H_0 (action is not necessary) if $\overline{x} \geq 52.6$. Reject H_0 (action is necessary) if $\overline{x} < 52.6$. "Action," in both cases, constitutes disowning the previous experiment, hence, rejecting the theory based on that experiment. The mean of the value of the dependent variable, $\overline{x}$, from thirty-two replications to be performed by the present experimenter, serves as the key in this decision.

19.4 Regression and Correlation

When two sets of variables are associated with any kind of relation, their relation can be represented on a graph as a set of points, each point determined by a pair of corresponding coordinates, one from each set. When there is a cause-effect relation, the values of the independent variable are shown on the horizontal axis and those of the dependent variable on the vertical axis. Such a graph is known as a *scatter diagram* (see Figure 19.1) because the points can be scattered without any obvious order. The statistical technique used to develop a mathematical equation representing the relation (if there is one) among the points is known as *regression analysis*. Depending on how closely such a mathematical equation represents the scatter among all the points, the strength of the relation between the two variables can be determined using a technique known as *correlation analysis*, the result of which is called the *correlation coefficient*.

19.4.1 Regression Analysis

The simplest mathematical relation among several points, each a relation between the independent and the dependent variable, is the equation of a straight line that best fits all the points of the scatter diagram. It is then said that the two variables have a *linear relation*. Figure 19.2 shows three variations of linear relation: (1) directly (or positively) related, (2) inversely (or negatively) related, and (3) not related, meaning there is no causal relation (though there is an association). We confine ourselves in this book to regression with linear relations only, though there are frequent cases wherein linear regression is not possible. For a study of nonlinear regression, the reader is advised to refer to an advanced book on statistics.

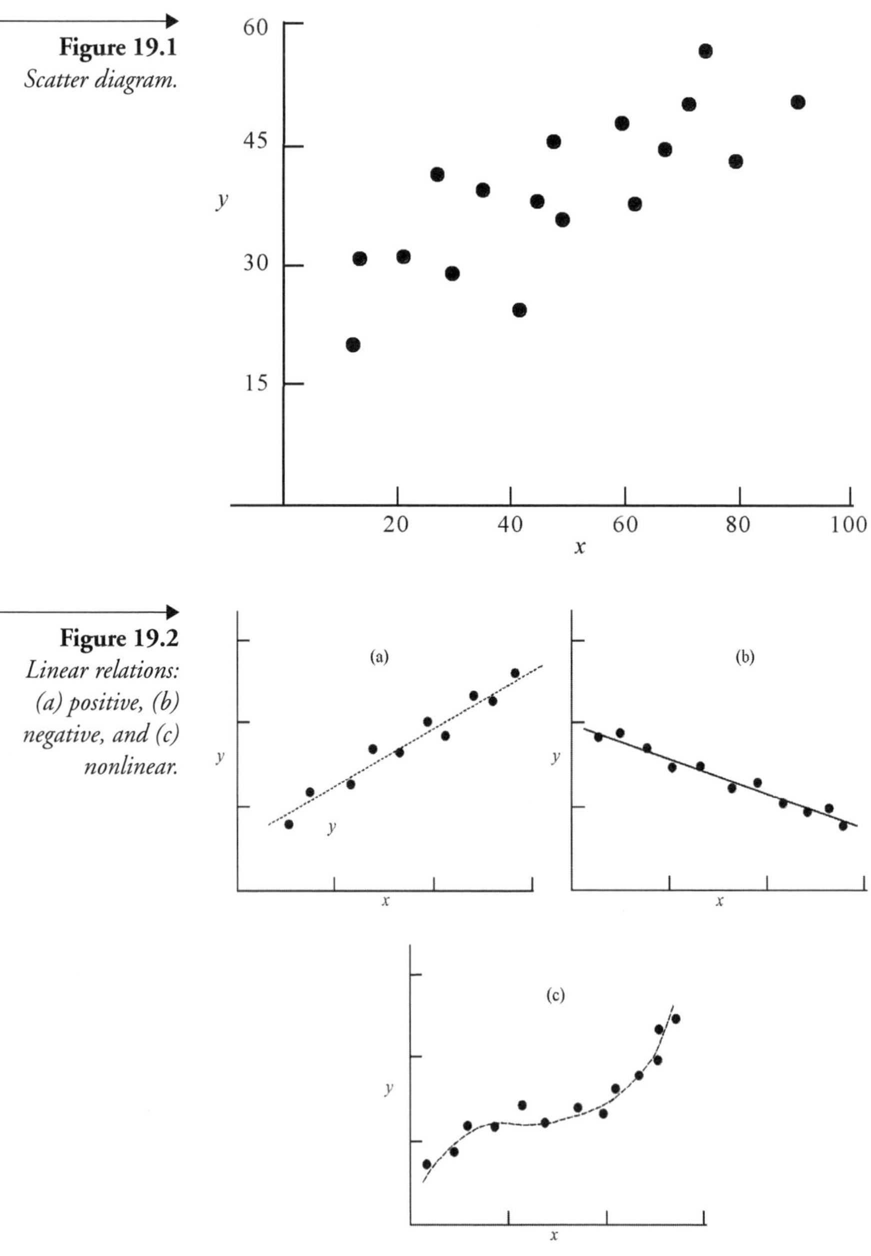

Figure 19.1
Scatter diagram.

Figure 19.2
Linear relations: (a) positive, (b) negative, and (c) nonlinear.

The straight line that best fits all the points of the scatter diagram is referred to as the *estimated regression line*, and the mathematical equation describing the line as the *estimated regression function*. When the function involves only one independent and one dependent variable, the analysis is referred to as a *simple linear regression*. The most commonly used and the best-suited simple linear regression of experimental results is known as the *least-square method*. The straight-line equation for this is expressed in the form

$$\ddot{y} = ax + p$$

where

$\ddot{y}$ = the estimated value of the dependent variable

x = the value of the independent variable

p = the intercept of the y-axis (value of $\ddot{y}$ when $x = 0$)

a = the change in the dependent variable corresponding to the change in the independent variable (also known as the *slope* of the line)

In this equation, constants a and p need to be evaluated independently. The value of a is given by

$$a = [N\Sigma x_i y_i - \Sigma x_i \, \Sigma y_i] \div [N\Sigma x_i^2 - (\Sigma x i)^2]$$

and that of p is given by

$$p = \bar{y} - a\bar{X}$$

where

x_i = the value of the independent variable for the ith observation (i.e., the point in the scatter diagram)

y_i = the value of the dependent variable corresponding to x_i

$\bar{X}$ = the mean value of the independent variable

$\bar{y}$ = the mean value of the dependent variable

N = the total number of observations (i.e., the number of points in the scatter diagram)

The sum of all the squares of the difference between the observed values of the dependent variable, y_i, and the estimated values of the dependent variable, $\ddot{y}$, meaning the vertical distances of these points from the regression line, will be a minimum when the regression line is drawn by this method. That is the reason this method is called the least-square method.

19.4.2 Measuring the Goodness of Regression

Once the regression line is drawn, it is possible, in principle, to find the y value for any given x value within the extremities; this, indeed, is the purpose of mathematizing the scatter diagram. But the regression line (determined by the regression function) itself is an approximation involving error because all of the points are not located on it but are scattered around it. The error involved is reduced to the least possible by means of regression; the quantity of the error, known as the *error sum of squares* (SSE), can be expressed as

$$\text{SSE} = \Sigma(y_i - \ddot{y}_i)^2 \tag{19.6}$$

Suppose the regression were not carried out and the mean of the y values of all the observations (points in the scatter diagram), $\bar{y}$, were used as the reference to express the deviation of all the points, we would have what is known as the *total sum of squares* (SST) about the mean, given by

$$\text{SST} = \Sigma(y_i - \bar{Y})^2 \tag{19.7}$$

which is also a measure of approximation, and the one with the most possible error in estimation. The benefit of carrying out the regression is that it reduces the possible error from the quantity given by SST to that given by SSE. The amount of such a benefit is referred to as the *regression sum of squares* (SSR), given in quantity by

$$\text{SSR} = \text{SST} - \text{SSE} \tag{19.8}$$

This relation is one of the most significant theorems of applied statistics. The ratio SSR ÷ SST is a measure of the benefit gained in doing the regression; known as *coefficient of determination*, it is represented by

$$r^2 = \text{SSR} \div \text{SST} \tag{19.9}$$

This ratio being always less than 1.0, the efficacy of the regression function can also be expressed as a percentage (of the total sum of squares).

Now looking back at (19.8), we want to point out that the value of SSR can be directly evaluated independently of SST and SSE, which can be easily "proved" by a numerical example. Leaving this to the reader (if confirmation is desired), the formula required is given below:

$$\text{SSR} = \Sigma(\ddot{y}_i - \overline{Y})^2 \tag{19.10}$$

where $\ddot{y}$ is the "estimated" y value of an observation (now to be read on the regression line), and $\overline{Y}$ is the mean of all the y values as observed. In passing, it may be mentioned that in the fields of business and economics, a correlation with $r^2 = 0.6$ or higher is considered satisfactory. In most laboratory experiments, a much higher r^2 than 0.6 is desirable. Figure 19.3 summarizes the methods for evaluating SSE, SST, and SSR, restricted for demonstration to only three data points.

Figure 19.3
Calculating SSE, SST, and SSR.

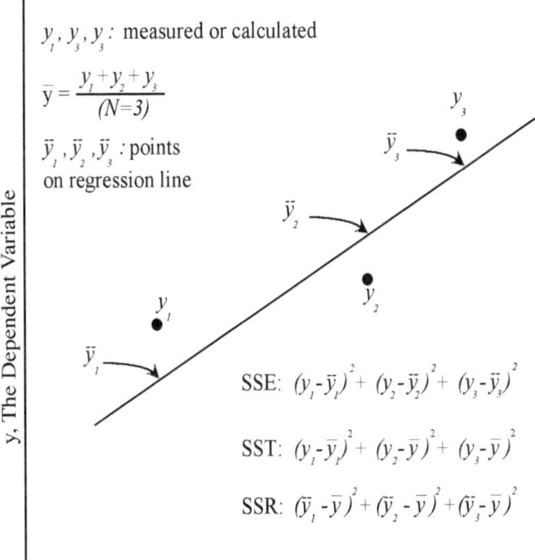

19.4.3 Correlation Coefficient

We remind ourselves that every set of *x-y* relations is a representative sample of a hypothetical population implied by generalization of the experimental observation. Quite often, the experimenter's main concern is to know whether or not x and y are related at all, and if they are, to know the strength of their relation. The statistical measure of such strength is known as the *correlation coefficient* and is given by the value

$$r = \pm \sqrt{r^2} \tag{19.11}$$

where r^2 is the *coefficient of determination.*

The way it is defined, r can have a "+" or "–" sign, the significance of which is as follows:

In the equation

$$y = ax + p,$$

if the slope (a) is positive, r is taken to be positive, and if (a) is negative, r is taken to be negative. The maximum value that r^2 can have being 1.0, r has the range +1 to –1. The correlation coefficient being either +1 or –1 means that the correlation is perfect, in turn meaning that all the points in the scatter diagram are already located on a straight line, a result that is seldom realized in experimental research.

Defining r as above may imply that regression analysis is a prerequisite to get the value of r. This is not so. A sample correlation coefficient can be calculated directly from the experimental observations, shown as the x and y coordinates of the points in the scatter diagram. The formula required is given below:

$$r = \frac{N\Sigma x_i y_i - \Sigma x_i \Sigma y_i}{\sqrt{N\Sigma x_i^2 - (\Sigma x_i)^2} \times \sqrt{N\Sigma y_i^2 - (\Sigma y_i)^2}} \tag{19.12}$$

where N is the sample size, that is, the number of observations made with different values of the independent variable, each observation shown as a point in the scatter diagram.

19.5 Multiple Regression

Thus far, we have dealt with simple linear regression, which is adequate for most lab experiments required in college course work. For research work, one-factor-at-a-time experiments are inadequate. It is now time for us to reflect on how close—or how far away—the reality of the experimental situation is to the results obtained by simple linear regression.

1. It has often been pointed out in this book that a single, independent variable acting alone on a dependent variable is quite an idealized situation; reality is far from it.
2. Because there is one independent variable affecting the dependent variable, we can represent their mutual relation as a number of points, each with two coordinates, x and y, in a two-dimensional space.
3. If more than one independent variable affects the dependent variable, we need, for representation on the same basis, three or more dimensions.
4. We made the assumption that the points representing the mutual relation of x and y are scattered around a straight line. To the extent that the points do not touch the straight line, we attribute the discrepancy to "experimental error" or to sacrifices made for simplification.
5. The straight line representing the relation of x and y is taken only as an estimate.
6. The reason for selecting the straight line, instead of a curve, which is more likely, is that we can use the polynomial of the first degree, thus avoiding the need to deal with an exponent of x that is either less or more than 1.0; this, in turn, makes the computation of the function of y easier. One could argue that thus restricting the exponent to 1.0 is not justified in view of the apparently unlimited computing capacity now available.

These considerations, besides others, serve as indicators of the limitations of linear regression. Even so, linear regressions are often useful for analyzing situations with two independent variables. The functional form of the equation, then, conforming to linear regression, is

$$\ddot{y} = ax_1 + bx_2 + p \tag{19.13}$$

In this form, the function is referred to as *multiple regression*, meaning two or more independent variables.

The solution of this function consists of finding the numerical values for *a*, *b*, and *p*, conforming to the following:

When different values of x_1 and x_2 are substituted, the corresponding $\ddot{y}$ values should represent the known values of the dependent variable. Further, and more importantly, the equation should become the instrument to estimate the value of the dependent variable, $\ddot{y}$, for any arbitrarily chosen values of x_1 and x_2, within the ranges of these variables as required in the experiment. The above equation can be solved considering *p* as the $\ddot{y}$ intercept and *a* and *b* as the slopes of the two straight lines, each conforming to the least-square method. The regression function will then be represented by a plane in a three-dimensional space, whose coordinates are $\ddot{y}$, x_1, and x_2. And, the plane will stretch between the two straight lines, somewhat like the fabric between two consecutive spokes of an umbrella. Solutions of such and more involved problems require the help of computer software programs, a host of which are now available, such as *Statistical Package for the Social Sciences* (*SPSS*), *Biomedical Computer Programs: P Series* (*BMDP*), and *Statistical Analysis Systems* (*SAS*). Considering (19.13), several combinations—usually a limited but sufficient number—of the numerical values of *y*, x_1, and x_2, in the required order, are input to the computer. The processed output typically contains the numerical values of *a*, *b*, and *p*, often supplemented with such information as the standard deviation and regression coefficient of the variables, the coefficient of determination, and the standard error of the estimation done.

Going another step further, we may think of the situation of three independent variables, x_1, x_2, and x_3, together causing effect on the dependent variable. Restricting the analysis to linear regression, we may state such a functional relation as

$$\ddot{y} = ax_1 + bx_2 + cx_3 + p. \tag{19.14}$$

In this case, we reach, in fact, go beyond the limits of linear regression. This is so because, in place of the plane with the previous situation of two independent variables, we now need a three-dimensional volume to represent the result of the linear regression. Even more formidable, we need to have a space of four dimensions in which such a volume should be placed.

The obvious question is, What do we resort to when we have three or more independent variables? For the answer and the applications, the reader is advised to revisit Chapter 9 of this book.

19.6 Bibliography

Anderson, David R., Dennis J. Sweeny, and Thomas A. Williams. *Introduction to Statistics: An Applications Approach.* St. Paul, MN: West Publishing Co., 1981.

Mandel, John. *The Statistical Analysis of Experimental Data.* Mineola, NY: Dover Publications, 1984.

Index